CIVIL ENGINEERING

CIVIL ENGINEERING DRAWING

BY

ALYNE COLE
LEVI MARTIN

KNOWLEDGE
BAKERS

KNOWLEDGE BAKERS

ISBN: 978-93-9001-365-4 (HB)

Published by : **Knowledge Bakers**
Pune, India
Email: info@knowledgebakers.com

PREFACE

Civil engineering drawing stands as a crucial cornerstone in the process of designing and constructing infrastructure that shapes our cities, towns, and landscapes. From the initial conceptualization of projects to the meticulous detailing required for implementation, accurate and comprehensive drawings serve as the language that bridges the gap between ideas and tangible structures. As we delve into the realm of civil engineering drawing, it becomes evident that proficiency in this art is not only essential for ensuring the functionality and safety of projects but also for fostering innovation and creativity in the field.

This book, titled "Civil Engineering Drawing" emerges as a comprehensive guide to understanding the principles, techniques, and applications of drawing in the civil engineering context. Through collaboration with experienced professionals and educators in the field, we aim to provide readers with a comprehensive overview of the tools, methodologies, and standards used in civil engineering drawing.

As editors of this volume, we have curated a collection of chapters that cover various aspects of civil engineering drawing, including technical drawing fundamentals, geometric construction, drafting conventions, dimensioning, detailing, and more. Each chapter offers practical examples, illustrative diagrams, and step-by-step tutorials to help readers develop proficiency in creating and interpreting engineering drawings accurately and efficiently.

Furthermore, this book seeks to highlight the importance of civil engineering drawing as a communication tool for conveying design intent, facilitating collaboration among multidisciplinary teams, and ensuring compliance with regulatory requirements. By showcasing real-world examples, best practices, and emerging trends, we aim to inspire readers to embrace drawing as an integral part of the civil engineering design process and harness its potential to drive innovation and excellence in the field.

As we embark on this journey of exploration and learning in the realm of civil engineering drawing, we extend our gratitude to the contributors whose expertise and dedication have enriched this volume. We hope that "Civil Engineering Drawing: Principles and Practices" serves as a valuable resource for students, professionals, and educators alike, empowering them to master the art of drawing and contribute to the advancement of civil engineering practice and knowledge.

Alyne Cole
Levi Martin

CONTENTS

1

ENGINEERING DRAWING: OVERVIEW

An engineering drawing is a subcategory of technical drawings. The purpose is to convey all the information necessary for manufacturing a product or a part. Engineering drawings use standardised language and symbols. This makes understanding the drawings simple with little to no personal interpretation possibilities.

An engineering drawing, a type of technical drawing, is created within the technical drawing discipline, and used to fully and clearly define requirements for engineered items. Engineering drawings are usually created in accordance with standardized conventions for layout, nomenclature, interpretation, appearance (such as typefaces and line styles), size, etc. One such standardized convention is called GD&T.

Each field in the Fields of engineering will have its own set of requirements for the producing drawings in terms line weight, symbols, and technical jargon. Some fields of engineering have no GD&T requirements.

The purpose of such a drawing is to accurately and unambiguously capture all the geometric features of a product or a component. The end goal of an engineering drawing is to convey all the required information that will allow a manufacturer to produce that component.

Engineering drawings used to be created by hand using tools such as pencils, ink, straightedges, T-squares, French curves, triangles, rulers, scales, and erasers. Today they are usually done electronically with computer-aided design (CAD).

The drawings are still often referred to as "blueprints" or "bluelines", although those terms are anachronistic from a literal perspective, since most copies of engineering drawings that were formerly made using a chemical-printing process that yielded graphics on blue-colored paper or, alternatively, of blue-lines on white paper, have been superseded by more modern reproduction processes that yield black or multicolour lines on white paper. The more generic term "print" is now in common usage in the U.S. to mean any paper copy of an engineering drawing.

The process of producing engineering drawings, and the skill of producing them, is often referred to as technical drawing or drafting, although technical drawings are also required for disciplines that would not ordinarily be thought of as parts of engineering.

THE PURPOSE OF ENGINEERING DRAWINGS

As already said, such a technical drawing has all the information for manufacturing a part or welding and building an assembly. The info includes dimensions, part names and numbers, etc. So once a manufacturing engineer gets the drawing, he can start the production process without a second thought.

First, we have to pause for a second and address our own customers here to avoid confusion. The drawings you submit for instant pricing and manufacturing in our system do not need any of this. The same applies to 3D models. CAD files and drawings made according to our design tips include all the necessary information for making your product. The only time we ask for a drawing is if you want to specify tolerances.

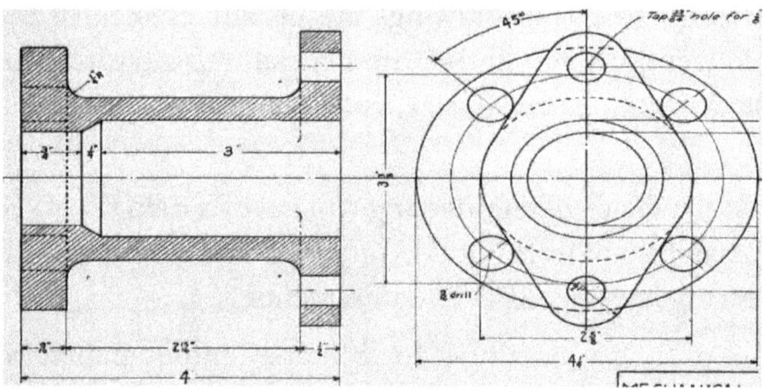

Fig 1.1

Still, knowing all the rules and basics of formatting is an absolute must in the industry, as traditional manufacturing companies still need detailed drawings.

How to Make Drawings?

A few decades ago, you would have had to sit down at a drawing board covered with papers of different size, rulers, callipers, etc. Today, all these instruments are still good for manual drafting but no contemporary manufacturer really wants such drawings.

Why? Because most of the machinery uses CNC systems that can read the information straight from the files and produce a cutting program accordingly. Drawings done by hand would just add a lot of manual work for manufacturing engineers.

HISTORY

Technical drawing has existed since ancient times. Complex technical drawings were made in renaissance times, such as the drawings of Leonardo da Vinci. Modern engineering drawing, with its precise conventions of orthographic projection and scale, arose in France at a time when the Industrial Revolution was in its infancy. L. T. C. Rolt's biography of Isambard Kingdom Brunel says of his father, Marc Isambard Brunel, that "It seems fairly certain that Marc's drawings of his block-making machinery (in 1799) made a contribution to British engineering technique much greater than the machines they represented. For it is safe to assume that he had mastered the art of presenting three-dimensional objects in a two-dimensional plane which we now call mechanical drawing. It had been evolved by Gaspard Monge of Mezieres in 1765 but had remained a military secret until 1794 and was therefore unknown in England."

Standardization and disambiguation

Engineering drawings specify requirements of a component or assembly which can be complicated. Standards provide rules for their specification and interpretation. Standardization also aids internationalization, because people from different countries who speak different languages can read the same engineering drawing, and interpret it the same way.

One major set of engineering drawing standards is ASME Y14.5 and Y14.5M (most recently revised in 2009). These apply widely in the United States, although ISO 8015 (Geometrical product specifications (GPS) — Fundamentals — Concepts, principles and rules) is now also important. In 2018, ASME AED-1 was created to develop advanced practices unique to aerospace and other industries and supplement to Y14.5 Standards.

In 2011, a new revision of ISO 8015 (Geometrical product specifications (GPS) — Fundamentals — Concepts, principles and rules) was published containing the Invocation Principle. This states that, "Once a portion of the ISO geometric product specification (GPS) system is invoked in a mechanical engineering product documentation, the entire ISO GPS system is invoked." It also goes on to state that marking a drawing "Tolerancing ISO 8015" is optional. The implication of this is that any drawing using ISO symbols can only be interpreted to ISO GPS rules. The only way not to invoke the ISO GPS system is to invoke a national or other standard. Britain, BS 8888 (Technical Product Specification) has undergone important updates in the 2010s.

COMMON FEATURES

Drawings convey the following critical information:

- **Geometry** - the shape of the object; represented as views; how the object will look when it is viewed from various angles, such as front, top, side, etc.
- **Dimensions** - the size of the object is captured in accepted units.
- **tolerances** - the allowable variations for each dimension.
- **Material** - represents what the item is made of.
- **Finish** - specifies the surface quality of the item, functional or cosmetic. For example, a mass-marketed product usually requires a much higher surface quality than, say, a component that goes inside industrial machinery.

LINE STYLES AND TYPES

A variety of line styles graphically represent physical objects. Types of lines include the following:

- **visible** - are continuous lines used to depict edges directly visible from a particular angle.
- **hidden** - are short-dashed lines that may be used to represent edges that are not directly visible.
- **center** - are alternately long- and short-dashed lines that may be used to represent the axes of circular features.
- **cutting plane** - are thin, medium-dashed lines, or thick alternately long- and double short-dashed that may be used to define sections for section views.
- **section** - are thin lines in a pattern (pattern determined by the material being "cut" or "sectioned") used to indicate surfaces in section views resulting from "cutting." Section lines are commonly referred to as "cross-hatching."
- **phantom** - (not shown) are alternately long- and double short-dashed thin lines used to represent a feature or component that is not part of the specified part or assembly. E.g. billet ends that may be used for testing, or the machined product that is the focus of a tooling drawing.

Lines can also be classified by a letter classification in which each line is given a letter.

- Type A lines show the outline of the feature of an object. They are the thickest lines on a drawing and done with a pencil softer than HB.
- Type B lines are dimension lines and are used for dimensioning, projecting, extending, or leaders. A harder pencil should be used, such as a 2H.
- Type C lines are used for breaks when the whole object is not shown. They are freehand drawn and only for short breaks. 2H pencil
- Type D lines are similar to Type C, except they are zigzagged and only for longer breaks. 2H pencil
- Type E lines indicate hidden outlines of internal features of an object. They are dotted lines. 2H pencil
- Type F lines are Type Ftypo lines, except they are used for drawings in electrotechnology. 2H pencil
- Type G lines are used for centre lines. They are dotted lines, but a long line of 10-20 mm, then a gap, then a small line of 2 mm. 2H pencil
- Type H lines are the same as Type G, except that every second long line is thicker. They indicate the cutting plane of an object. 2H pencil
- Type K lines indicate the alternate positions of an object and the line taken by that object. They are drawn with a long line of 10-20 mm, then a small gap, then a small line of 2 mm, then a gap, then another small line. 2H pencil.

Orthographic projection

The orthographic projection shows the object as it looks from the front, right, left, top, bottom, or back, and are typically positioned relative to each other according to the rules of either first-angle or third-angle projection.

- First angle projection is the ISO standard and is primarily used in Europe. The 3D object is projected into 2D "paper" space as if you were looking at an X-ray of the object: the top view is under the front view, the right view is at the left of the front view.
- Third angle projection is primarily used in the United States and Canada, where it is the default projection system according to BS 8888:2006, the left view is placed on the left and the top view on the top.

Not all views are necessarily used, and determination of what surface constitutes the front, back, top and bottom varies depending on the projection used.

Auxiliary projection

An auxiliary view is an orthographic view that is projected into any plane other than one of the six principal views. These views are typically used when an object contains some sort of inclined plane. Using the auxiliary view allows for that inclined plane (and any other significant features) to be projected in their true size and shape. The true size and shape of any feature in an engineering drawing can only be known when the Line of Sight (LOS) is perpendicular to the plane being referenced.

Isometric projection

The isometric projection show the object from angles in which the scales along each axis of the object are equal. Isometric projection corresponds to rotation of the object by ± 45° about the vertical axis, followed by rotation of approximately ± 35.264° [= arcsin(tan(30°))] about the horizontal axis starting from an orthographic projection view. "Isometric" comes from the Greek for "same measure". One of the things that makes isometric drawings so attractive is the ease with which 60 degree angles can be constructed with only a compass and straightedge.

Isometric projection is a type of axonometric projection. The other two types of axonometric projection are:

- Dimetric projection
- Trimetric projection

Oblique projection

An oblique projection is a simple type of graphical projection used for producing pictorial, two-dimensional images of three-dimensional objects:

- it projects an image by intersecting parallel rays (projectors)
- from the three-dimensional source object with the drawing surface (projection plan).

In both oblique projection and orthographic projection, parallel lines of the source object produce parallel lines in the projected image.

Perspective

Perspective is an approximate representation on a flat surface, of an image as it is perceived by the eye. The two most characteristic features of perspective are that objects are drawn:

- Smaller as their distance from the observer increases
- Foreshortened: the size of an object's dimensions along the line of sight are relatively shorter than dimensions across the line of sight.

Section Views

Projected views (either Auxiliary or Orthographic) which show a cross section of the source object along the specified cut plane. These views are commonly used to show internal features with more clarity than may be available using regular projections or hidden lines. In assembly drawings, hardware components (e.g. nuts, screws, washers) are typically not sectioned.

Basic Components of an Engineering Drawing

Let's see what makes up an engineering drawing. A single drawing includes many elements with quite a few variations to each of them. So let's take a closer look here.

DIFFERENT TYPES OF LINES

Not every line on an engineering drawing is equal. The different options make it possible to show both visible and hidden edges of a part, centre lines, etc.

Fig 1.2

The most common is a continuous line, also known as a drawing line. This represents the physical boundaries of an object. Put simply, these lines are for drawing the objects. The line thickness varies – the outer contour uses thicker lines and inner lines are thinner.

Hidden lines can show something that would not be otherwise visible on the drawings. For example, hidden lines may show the length of an internal step in a turned part without using a section or a cutout view (we explain both later).

Centre lines are used to show hole and the symmetric properties of parts. Showing symmetricity can reduce the number of dimensions and make the drawing more eye-pleasing, thus easier to read.

Fig 1.3

Extension lines annotate what is being measured. The dimension line has two arrowheads between the extension lines and the measurement on top (or inside, like in the image above) the line.

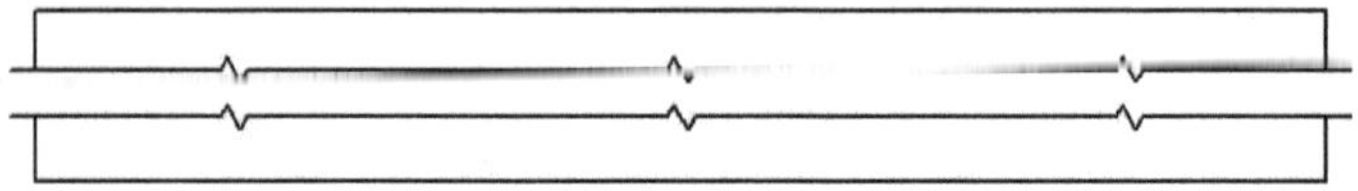

Fig 1.4

Break lines indicate that a view has been broken. If you have a part that is 3000 mm long and 10 mm wide with symmetric properties, using a break-out makes gives all the info without using as much space.

While a good way for giving information to people, CNC machines need full views in order to cut the parts. Otherwise, the manufacturing engineer has to reconstruct the whole part from the measurements.

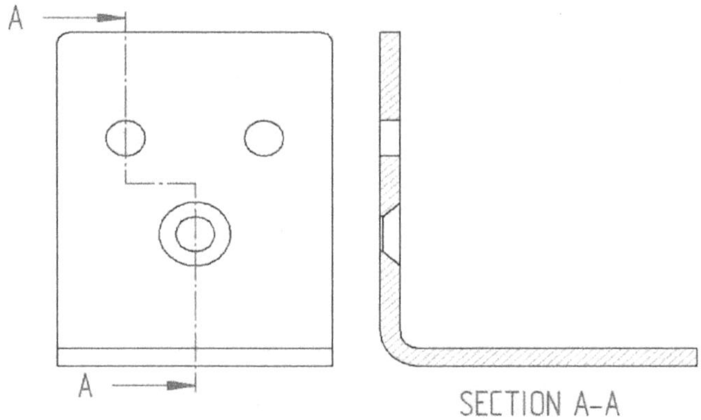

Fig 1.5

When using a cutout view, the cutting plane lines show the trajectory of the cutout. Here you can see that the A-A cutting line brings both types of holes into the view.

Types of Views

So let's take a closer look at the different types of views that are often present in a manufacturing drawing. Each serves a certain purpose. Bear in mind that adding views should follow the same logic as dimensioning – include as little as possible and as much as necessary.

A tip for good engineering practice – only include a view if it contributes to the overall understanding of the design.

Isometric View

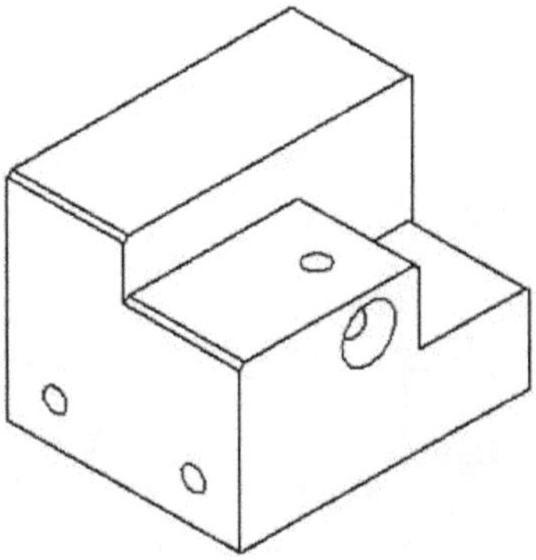

Fig 1.6

Isometric drawings show parts as three-dimensional. All the vertical lines stay vertical (compared to front view) and otherwise parallel lines are shown on a 30-degree angle.

The lines that are vertical and parallel are in their true length. Which means you can use a ruler and the scaling of the drawing to easily measure the length straight from a paper drawing, for example. The same does not apply to angled lines.

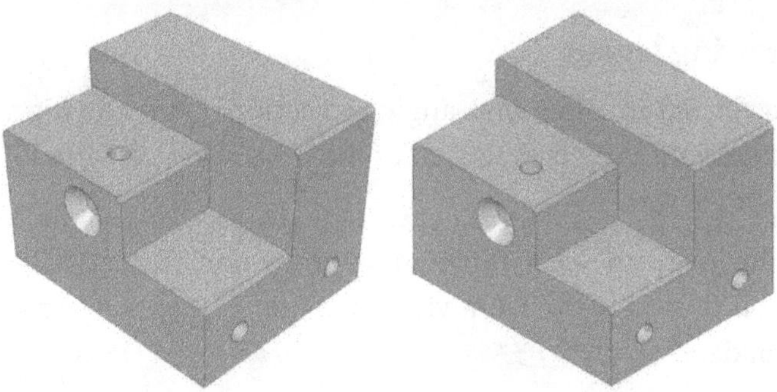

Fig 1.7

Left – perspective; right – isometric

It is important to distinguish the isometric view from a perspective view. A perspective view is an artistic one that represents an object as it seems to the eye. Engineers stay true to the dimensions rather than optical illusions.

Orthographic View

This is the bread and butter of an engineering drawing. An orthographic view or orthographic projection is a way of representing a 3D object in 2 dimensions.

Thus, a 2D view has to convey everything necessary for part production. This kind of representation allows avoiding any kind of distortion of lengths.

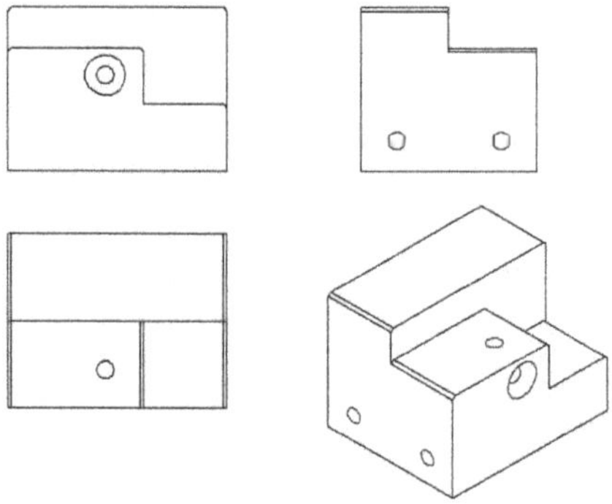

Fig 1.8

The most common way to communicate all the information is by using three different views in a multiview drawing:

It may be possible that some additional views are necessary to show all the info. But again, less is more.

The positioning of the views differs a bit regionally. For example, look at the image below to compare the US and ISO layouts.

ISO vs US drawing layout

The one on the left is called first-angle projection. Here, the top view is under the front view, the right view is at the left of the front view, etc. The ISO standard is primarily used in Europe.

On the right, you can see a third-angle projection. The right view is on the right, top view on the top of the front view, etc. This system is especially popular in the US and Canada.

Flat Pattern

Flat pattern drawing

If you are making a folded sheet metal part, do not forget to add a flat pattern view. The cutting job comes before bending. When it comes to our customers, the easiest way is just to upload a STEP file without any accompanying drawings.

Systems of dimensioning and tolerancing

Almost all engineering drawings (except perhaps reference-only views or initial sketches) communicate not only geometry (shape and location) but also dimensions and tolerances for those characteristics. Several systems of dimensioning and tolerancing have evolved. The simplest dimensioning system just specifies distances between points (such as an object's length or width, or hole center locations). Since the advent of well-developed interchangeable manufacture, these distances have been accompanied by tolerances of the plus-or-minus or min-and-max-limit types. Coordinate dimensioning involves defining all points, lines, planes, and profiles in terms of Cartesian coordinates, with a common origin. Coordinate dimensioning was the sole best option until the post-World War II era saw the development of geometric dimensioning and tolerancing (GD&T), which departs from the limitations of coordinate dimensioning (e.g., rectangular-only tolerance zones, tolerance stacking) to allow the most logical tolerancing of both geometry and dimensions (that is, both form [shapes/locations] and sizes).

CIVIL ENGINEERING

It is a professional engineering discipline that deals with the design, construction, and maintenance of the physical and naturally built environment, including works like bridges, roads, canals, dams, and buildings. Civil engineering is the oldest engineering discipline after military engineering, and it was defined to distinguish non-military engineering from military engineering.

It is traditionally broken into several sub-disciplines including environmental engineering, geotechnical engineering, structural engineering, transportation engineering, municipal or urban engineering, water resources engineering, materials engineering, coastal engineering, surveying, and construction engineering. Civil engineering takes place on all levels: in the public sector from municipal through to national governments, and in the private sector from individual homeowners through to international companies.

Engineering has been an aspect of life since the beginnings of human existence. The earliest practices of Civil engineering may have commenced between 4000 and 2000 BC in Ancient Egypt and Mesopotamia (Ancient Iraq) when humans started to abandon a nomadic existence, thus causing a need for the construction of shelter. During this time, transportation became increasingly important leading to the development of the wheel and sailing.

Until modern times there was no clear distinction between civil engineering and architecture, and the term engineer and architect were mainly geographical variations referring to the same person, often used interchangeably. The construction of Pyramids in Egypt (circa 2700-2500 BC) might be considered the first instances of large structure constructions. Other ancient historic civil engineering constructions include the Parthenon by Iktinos in Ancient Greece (447-438 BC), the Appian Way by Roman engineers (c. 312 BC), the Great Wall of China by General Meng T'ien under orders from Ch'in Emperor Shih Huang Ti (c. 220 BC) and the stupas constructed in ancient Sri Lanka like the Jetavanaramaya and the extensive irrigation works in Anuradhapura. The Romans developed civil structures throughout their empire, including especially aqueducts, insulae, harbours, bridges, dams and roads.

In the 18th century, the term civil engineering was coined to incorporate all things civilian as opposed to military engineering. The first self-proclaimed civil engineer was John Smeaton who constructed the Eddystone Lighthouse. In 1771 Smeaton and some of his colleagues formed the Smeatonian Society of Civil Engineers, a group of leaders of the profession who met informally over dinner. Though there was evidence of some technical meetings, it was little more than a social society.

In 1818 the Institution of Civil Engineers was founded in London, and in 1820 the eminent engineer Thomas Telford became its first president. The institution received a Royal Charter in 1828, formally recognising civil engineering as a profession. Its charter defined civil engineering as:

the art of directing the great sources of power in nature for the use and convenience of man, as the means of production and of traffic in states, both for external and internal trade, as applied in the construction of roads, bridges, aqueducts, canals, river navigation and docks for internal intercourse and exchange, and in the construction of ports, harbours, moles, breakwaters and lighthouses, and in the art of navigation by artificial power for the purposes of commerce, and in the construction and application of machinery, and in the drainage of cities and towns.

The first private college to teach Civil Engineering in the United States was Norwich University founded in 1819 by Captain Alden Partridge. The first degree in Civil Engineering in the United States was awarded by Rensselaer Polytechnic Institute in 1835. The first such degree to be awarded to a woman was granted by Cornell University to Nora Stanton Blatch in 1905.

Civil engineering is the application of physical and scientific principles, and its history is intricately linked to advances in understanding of physics and mathematics throughout history. Because civil engineering is a wide ranging profession, including several separate specialized sub-disciplines, its history is linked to knowledge of structures, materials science, geography, geology, soils, hydrology, environment, mechanics and other fields.

Throughout ancient and medieval history most architectural design and construction was carried out by artisans, such as stone masons and carpenters, rising to the role of master builder. Knowledge was retained in guilds and seldom supplanted by advances. Structures, roads and infrastructure that existed were repetitive, and increases in scale were incremental.

One of the earliest examples of a scientific approach to physical and mathematical problems applicable to civil engineering is the work of Archimedes in the 3rd century BC, including Archimedes Principle, which underpins our understanding of buoyancy, and practical solutions such as Archimedes' screw. Brahmagupta, an Indian mathematician, used arithmetic in the 7th century AD, based on Hindu-Arabic numerals, for excavation (volume) computations.

The civil engineer

Education and licensure

Civil engineers typically possess an academic degree with a major in civil engineering. The length of study for such a degree is usually three to five years and the completed degree is usually designated as a Bachelor of Engineering, though some universities designate the degree as a Bachelor of Science. The degree generally includes units covering physics, mathematics, project management, design and specific topics in civil engineering. Initially such topics cover most, if not all, of the sub-disciplines of civil engineering.

Students then choose to specialize in one or more sub-disciplines towards the end of the degree. While an Undergraduate (BEng/BSc) Degree will normally provide successful students with industry accredited qualification, some universities offer postgraduate engineering awards (MEng/MSc) which allow students to further specialize in their particular area of interest within engineering.

In most countries, a Bachelor's degree in engineering represents the first step towards professional certification and the degree program itself is certified by a professional body. After completing a certified degree program the engineer must satisfy a range of requirements (including work experience and exam requirements) before being certified. Once certified, the engineer is designated the title of Professional Engineer (in the United States, Canada and South Africa), Chartered Engineer (in most Commonwealth countries), Chartered Professional Engineer (in Australia and New Zealand), or European Engineer (in much of the European Union). There are international engineering agreements between relevant professional bodies which are designed to allow engineers to practice across international borders.

The advantages of certification vary depending upon location. For example, in the United States and Canada "only a licensed engineer may prepare, sign and seal, and submit engineering plans and drawings to a public authority for approval, or seal engineering work for public and private clients.". This requirement is enforced by state and provincial legislation such as Quebec's Engineers Act. In other countries, no such legislation exists. In Australia, state licensing of engineers is limited to the state of Queensland. Practically all certifying bodies maintain a code of ethics that they expect all members to abide by or risk expulsion.

In this way, these organizations play an important role in maintaining ethical standards for the profession. Even in jurisdictions where certification has little or no legal bearing on work, engineers are subject to contract law. In cases where an engineer's work fails he or she may be subject to the tort of negligence and, in extreme cases, the charge of criminal negligence. An engineer's work must also comply with numerous other rules and regulations such as building codes and legislation pertaining to environmental law.

CAREERS

There is no one typical career path for civil engineers. Most people who graduate with civil engineering degrees start with jobs that require a low level of responsibility, and as the new engineers prove their competence, they are trusted with tasks that have larger consequences and require a higher level of responsibility. However, within each branch of civil engineering career path options vary.

In some fields and firms, entry-level engineers are put to work primarily monitoring construction in the field, serving as the "eyes and ears" of senior design engineers; while in other areas, entry-level engineers perform the more routine tasks of analysis or design and interpretation. Experienced engineers generally do more complex analysis or design work, or management of more complex design projects, or management of other engineers, or into specialized consulting, including forensic engineering.

Sub-disciplines

In general, civil engineering is concerned with the overall interface of human created fixed projects with the greater world. General civil engineers work closely with surveyors and specialized civil engineers to fit and serve fixed projects within their given site, community and terrain by designing grading, drainage, pavement, water supply, sewer service, electric and communications supply, and land divisions.

General engineers spend much of their time visiting project sites, developing community consensus, and preparing construction plans. General civil engineering is also referred to as site engineering, a branch of civil engineering that primarily focuses on converting a tract of land from one usage to another. Civil engineers typically apply the principles of geotechnical engineering, structural engineering, environmental engineering, transportation engineering and construction engineering to residential, commercial, industrial and public works projects of all sizes and levels of construction.

Coastal engineering

Coastal engineering is concerned with managing coastal areas. In some jurisdictions the terms sea defense and coastal protection are used to mean, respectively, defence against flooding and erosion. The term coastal defence is the more traditional term, but coastal management has become more popular as the field has expanded to include techniques that allow erosion to claim land.

Construction engineering

Construction engineering involves planning and execution of the designs from transportation, site development, hydraulic, environmental, structural and geotechnical engineers. As construction firms tend to have higher business risk than other types of civil engineering firms, many construction engineers tend to take on a role that is more business-like in nature: drafting and reviewing contracts, evaluating logistical operations, and closely-monitoring prices of necessary supplies.

ENVIRONMENTAL ENGINEERING

Environmental engineering deals with the treatment of chemical, biological, and/or thermal waste, the purification of water and air, and the remediation of contaminated sites, due to prior waste disposal or accidental contamination. Among the topics covered by environmental engineering are pollutant transport, water purification, waste water treatment, air pollution, solid waste treatment and hazardous waste management.

Environmental engineers can be involved with pollution reduction, green engineering, and industrial ecology. Environmental engineering also deals with the gathering of information on the environmental consequences of proposed actions and the assessment of effects of proposed actions for the purpose of assisting society and policy makers in the decision making process.

Environmental engineering is the contemporary term for sanitary engineering, though sanitary engineering traditionally had not included much of the hazardous waste management and environmental remediation work covered by the term environmental engineering. Some other terms in use are public health engineering and environmental health engineering.

Geotechnical engineering

Geotechnical engineering is an area of civil engineering concerned with the rock and soil that civil engineering systems are supported by. Knowledge from the fields of geology, material science and testing, mechanics, and hydraulics are applied by geotechnical engineers to safely and economically design foundations, retaining walls, and similar structures. Environmental concerns in relation to groundwater and waste disposal have spawned a new area of study called geoenvironmental engineering where biology and chemistry are important.

Some of the unique difficulties of geotechnical engineering are the result of the variability and properties of soil. Boundary conditions are often well defined in other branches of civil engineering, but with soil, clearly defining these conditions can be impossible. The material properties and behavior of soil are also difficult to predict due to the variability of soil and limited investigation.

This contrasts with the relatively well defined material properties of steel and concrete used in other areas of civil engineering. Soil mechanics, which describes the behavior of soil, is also complicated because soils exhibit nonlinear (stress-dependent) strength, stiffness, and dilatancy (volume change associated with application of shear stress).

Water resources engineering

Water resources engineering is concerned with the collection and management of water (as a natural resource). As a discipline it therefore combines hydrology, environmental science, meteorology, geology, conservation, and resource management. This area of civil engineering relates to the prediction and management of both the quality and the quantity of water in both underground (aquifers) and above ground (lakes, rivers, and streams) resources.

Water resource engineers analyze and model very small to very large areas of the earth to predict the amount and content of water as it flows into, through, or out of a facility. Although the actual design of the facility may be left to other engineers.

Hydraulic engineering is concerned with the flow and conveyance of fluids, principally water. This area of civil engineering is intimately related to the design of pipelines, water supply network, drainage facilities (including bridges, dams, channels, culverts, levees, storm sewers), and canals. Hydraulic engineers design these facilities using the concepts of fluid pressure, fluid statics, fluid dynamics, and hydraulics, among others.

MATERIALS ENGINEERING

Another aspect of Civil engineering is materials science. Material engineering deals with ceramics such as concrete, mix asphalt concrete, metals Focus around increased strength, metals such as aluminum and steel, and polymers such as polymethylmethacrylate (PMMA) and carbon fibers.

Materials engineering also consists of protection and prevention like paints and finishes. Alloying is another aspect of material engineering, combining two different types of metals to produce a stronger metal.

Structural engineering

Structural engineering is concerned with the structural design and structural analysis of buildings, bridges, towers, flyovers, tunnels, off shore structures like oil and gas fields in the sea, and other structures. This involves identifying the loads which act upon a structure and the forces and stresses which arise within that structure due to those loads, and then designing the structure to successfully support and resist those loads.

The loads can be self weight of the structures, other dead load, live loads, moving (wheel) load, wind load, earthquake load, load from temperature change etc. The structural engineer must design structures to be safe for their users and to successfully fulfill the function they are designed for (to be serviceable). Due to the nature of some loading conditions, sub-disciplines within structural engineering have emerged, including wind engineering and earthquake engineering.

Design considerations will include strength, stiffness, and stability of the structure when subjected to loads which may be static, such as furniture or self-weight, or dynamic, such as wind, seismic, crowd or vehicle loads, or transitory, such as temporary construction loads or impact. Other considerations include cost, constructability, safety, aesthetics and sustainability.

Surveying

Surveying is the process by which a surveyor measures certain dimensions that generally occur on the surface of the Earth. Surveying equipment, such as levels and theodolites, are used for accurate measurement of angular deviation, horizontal, vertical and slope distances. With computerisation, electronic distance measurement (EDM), total stations, GPS surveying and laser scanning have supplemented (and to a large extent supplanted) the traditional optical instruments. This information is crucial to convert the data into a graphical representation of the Earth's surface, in the form of a map.

This information is then used by civil engineers, contractors and even realtors to design from, build on, and trade, respectively. Elements of a building or structure must be correctly sized and positioned in relation to each other and to site boundaries and adjacent structures. Although surveying is a distinct profession with separate qualifications and licensing arrangements, civil engineers are trained in the basics of surveying and mapping, as well as geographic information systems. Surveyors may also lay out the routes of railways, tramway tracks, highways, roads, pipelines and streets as well as position other infrastructures, such as harbors, before construction.

LAND SURVEYING

In the United States, Canada, the United Kingdom and most Commonwealth countries land surveying is considered to be a distinct profession. Land surveyors are not considered to be engineers, and have their own professional associations and licencing requirements. The services of a licenced land surveyor are generally required for boundary surveys (to establish the boundaries of a parcel using its legal description) and subdivision plans (a plot or map based on a survey of a parcel of land, with boundary lines drawn inside the larger parcel to indicated the creation of new boundary lines and roads), both of which are generally referred to as cadastral surveying.

Construction Surveying

Construction surveying is generally performed by specialised technicians. Unlike land surveyors, the resulting plan does not have legal status. Construction surveyors perform the following tasks:

- Survey existing conditions of the future work site, including topography, existing buildings and infrastructure, and even including underground infrastructure whenever possible;
- Construction surveying (otherwise "lay-out" or "setting-out"): to stake out reference points and markers that will guide the construction of new structures such as roads or buildings for subsequent construction;
- Verify the location of structures during construction;
- As-Built surveying: a survey conducted at the end of the construction project to verify that the work authorized was completed to the specifications set on plans.

Transportation engineering

Transportation engineering is concerned with moving people and goods efficiently, safely, and in a manner conducive to a vibrant community. This involves specifying, designing, constructing, and maintaining transportation infrastructure which includes streets, canals, highways, rail systems, airports, ports, and mass transit.

It includes areas such as transportation design, transportation planning, traffic engineering, some aspects of urban engineering, queueing theory, pavement engineering, Intelligent Transportation System (ITS), and infrastructure management.

MUNICIPAL OR URBAN ENGINEERING

Municipal engineering is concerned with municipal infrastructure. This involves specifying, designing, constructing, and maintaining streets, sidewalks, water supply networks, sewers, street lighting, municipal solid waste management and disposal, storage depots for various bulk materials used for maintenance and public works (salt, sand, etc.), public parks and bicycle paths.

In the case of underground utility networks, it may also include the civil portion (conduits and access chambers) of the local distribution networks of electrical and telecommunications services. It can also include the optimizing of waste collection and bus service networks. Some of these disciplines overlap with other civil engineering specialties, however municipal engineering focuses on the coordination of these infrastructure networks and services, as they are often built simultaneously, and managed by the same municipal authority.

Architect's scale

An architect's scale is a specialized ruler. It is used in making or measuring from reduced scale drawings, such as blueprints and floor plans. It is marked with a range of calibrated scales (ratios). For accuracy and longevity the material used should be dimensionally stable and durable. Scales were traditionally made of wood, but today they are usually made of rigid plastic or aluminium. Architect's scales may be flat, with 4 scales, or have a symmetric 3-lobed cross-section, with 12 scales.

Architect's scale rulers used in Britain and other metric countries are marked with ratios without reference to a base unit. Therefore a drawing will indicate both its scale (ratio) and the unit of measurement being used.

In Britain, and elsewhere, the standard units used on architectural drawings are the (SI) units millimetres (mm) and metres (m), whereas in France centimetres (cm) and metres are most often used.

In Britain, for flat rulers, the paired scales often found on architect's scales are:

- 1:1/1:100
- 1:5/1:50

- 1:10/1:100
- 1:20/1:200
- 1:1250/1:2500

For triangular rulers, the paired scales are:

- 1:1/1:10
- 1:2/1:20
- 1:5/1:50
- 1:100/1:200
- 1:500/1:1000
- 1:1250/1:2500

Less common scales are:

- 1:25/1:250
- 1:331?3

United States and Imperial units

In the United States, and prior to metrification in Britain, Canada and Australia, architect's scales are/were marked as a ratio of x inches-to-the-foot (typically written as x"=1'-0"). For example one inch measured from a drawing with a scale of "one-inch-to-the-foot" is equivalent to one foot in the real world (a scale of 1:12) whereas one inch measured from a drawing with a scale of "two-inches-to-the-foot" is equivalent to six inches in the real world (a scale of 1:6). It is not to be confused with a true unitless ratio (inches to inches)-- a 1:50 architectural scale would be a 1:60 unitless scale.

Typical scales used in the United States are:

- Full scale, with inches divided into sixteenths of an inch

The following scales are generally grouped in pairs using the same dual-numbered index line (one scale is read from the right, and the other scale is read from the left):

- three-inches-to-the-foot (3"=1'-0") (ratio equivalent 1:4)/one-and-one-half-inch-to-the-foot (1½"=1'-0") (1:8)
- one-inch-to-the-foot (1"=1'-0") (1:12)/one-half-inch-to-the-foot (1/2"=1'-0") (1:24)
- three-quarters-inch-to-the-foot (3/4"=1'-0") (1:16)/three-eighths-inch-to-the-foot (3/8"=1'-0") (1:32)

- one-quarter-inch-to-the-foot (1/4"=1'-0") (1:48)/one-eighth-inch-to-the-foot (1/8"=1'-0") (1:96)
- three-sixteenths-inch-to-the-foot (3/16"=1'-0") (1:64)/three-thirty-seconds-inch-to-the-foot (3/32"=1'0") (1:128)

Plan (drawing)

Plans are a set of two-dimensional diagrams or drawings used to describe a place or object, or to communicate building or fabrication instructions. Usually plans are drawn or printed on paper, but they can take the form of a digital file.

Plans are often for technical purposes such as architecture, engineering, or planning. Their purpose in these disciplines is to accurately and unambiguously capture all the geometric features of a site, building, product or component. Plans can also be for presentation or orientation purposes, and as such are often less detailed versions of the former. The end goal of plans is either to portray an existing place or object, or to convey enough information to allow a builder or manufacturer to realize a design.

The term "plan" may casually be used to refer to a single view, sheet, or drawing in a set of plans. More accurately, plan refers to an orthographic projection looking down on the object, such as in a plan view or floor plan. The process of producing plans, and the skill of producing them, is often referred to as technical drawing.

Plan features

Format

Plans are often prepared in a "set". The set includes all the information required for the purpose of the set, and may exclude views or projections which are unnecessary. A set of plans can be on standard office-sized paper or on large sheets. It can be stapled, folded or rolled as required. A set of plans can also take the form of a digital file in a proprietary format such as DWG or an exchange file format such as DXF or PDF.

Plans are often referred to as "blueprints" or "bluelines". However, the terms are rapidly becoming an anachronism, since most copies of plans that were formerly made using a chemical-printing process that yielded graphics on blue-colored paper or, alternatively, of blue-lines on white paper, have been superseded by more modern reproduction processes that yield black or multicolour lines on white paper.

Scale

Plans are usually "scale drawings", meaning that the plans are drawn at specific ratio relative to the actual size of the place or object. Various scales may be used for different drawings in a set. For example, a floor plan may be drawn at 1:48 (or 1/4”=1’-0”) whereas a detailed view may be drawn at 1:24 (or 1/2”=1’-0”). Site plans are often drawn at 1” = 20’ (1:240) or 1” = 30’ (1:360).

In the metric system the ratios commonly are 1:5, 1:10, 1:20, 1:50, 1:100, 1:200, 1:500, 1:1000, 1:2000 and 1:5000

VIEWS AND PROJECTIONS

Because plans represent three-dimensional objects on a two-dimensional plane, the use of views or projections is crucial to the legibility of plans. Each projection is achieved by assuming a vantage point from which to see the place or object, and a type of projection. These projection types are:

- Parallel projection
  - Orthographic projection, including:
    - Plan view or floor plan view
    - Elevation, usually a side view of an exterior
    - Section , a view of the interior at a particular cutting plane
  - Axonometric projection, including:
    - Isometric projection
    - Dimetric projection
    - Trimetric projection
  - Oblique projection, and
- Perspective projection

Planning approach

There is no universal standard for sheet order, however the following describes a common approach:

- **General Information :** The first sheets in a set may include notes, assembly descriptions, a rendering of the project, or simply the project title.
- **Site :** Site plans, including a key plan, appear before other plans and on smaller projects may be on the first sheet. A projects could require a landscape plan, although this can be integrated with the site plan if the drawing remains clear.
- **Specific plans :** Floor plans, starting with the lowest floor and ending with the roof plan usually appear near the beginning of the set. Further,

for example, reflected Ceiling Plans (RCP)s showing ceiling layouts appear after the floor plans.

- **Elevations :** Starting with the principal, or front elevation, all the building elevations appear after the plans. Smaller residential projects may display the elevations before the plans. Elevation details may appear on the same sheets as the building elevations.
- **Sections:** Building sections that describe views cut through the entire building appear next, followed by wall sections, then detail sections.
- **Details:** Details may appear on any of the previous sheets, or may be collected to appear on detail sheets. These details may include construction details that show how the components of the building fit together. These details may also include millwork drawings or other interior details.
- **Schedules:** Many aspects of a building must be scheduled on larger projects. These include schedules for windows, doors, hardware, landscaping elements, rooms, and areas.

More aspects are, for example:

- **Structural:** While smaller projects may only show structural information on the plans and sections, larger projects have separate sheets describing the structure of the building.
- **Mechanical:** Mechanical drawings show plumbing and heating systems.
- **Electrical :** Again, many electrical layouts are shown on the floor plans, however larger projects have this information on separate sheets.

Architectural Plan

An architectural plan is a plan for architecture, and the documentation of written and graphic descriptions of the architectural elements of a building project including sketches, drawings and details.

The term "Architectural plan" can have multiple related meanings:

- Plan for an architectural project
- Documentation of written and graphic descriptions of the architectural elements of a building project including sketches, drawings and details. This effort could also includes both the design of new buildings and other structures, as well as the planning for reconstruction of early historic structures.
- Architectural design

- Floor plan
- Scale drawing of a structure. "the plans for City Hall were on file".

This article will focus on the general meaning of architectural plan as a plan and documentation for a building project.

An architectural plan is a type of plan for architecture to describe a place or object, or to communicate building or fabrication instructions. It can containing technical architectural drawings and documentation, and are drawn or printed on paper, but they can take the form of a digital file.

Their purpose in these disciplines is to accurately and unambiguously capture all the geometric features of a site, building, product or component. Plans can also be for presentation or orientation purposes, and as such are often less detailed versions of the former. The end goal of plans is either to portray an existing place or object, or to convey enough information to allow a builder or manufacturer to realize a design.

Types of architectural plans

Floor plan

A floor plan in architecture and building engineering is a diagram, usually to scale, of the relationships between rooms, spaces and other physical features at one level of a structure. Dimensions are usually drawn between the walls to specify room sizes and wall lengths. Floor plans will also include details of fixtures like sinks, water heaters, furnaces, etc. Floor plans will include notes to specify finishes, construction methods, or symbols for electrical items.

Similar to a map in an floor plan the orientation of the view is downward from above, but unlike a conventional map, a plan is understood to be drawn at a particular vertical position (commonly at about 3 feet above the floor). Objects below this level are seen, objects at this level are shown 'cut' in plan-section, and objects above this vertical position within the structure are omitted or shown dashed. Plan view or "planform" is defined as a vertical orthographic projection of an object on a horizontal plane, like a map.

Garden design

Garden design is the art and process of designing and creating plans for layout and planting of gardens and landscapes. Garden design may be done by the garden owner themselves, or by professionals of varying levels of experience and expertise. Most professional garden designers are trained in principles of design and in horticulture, and have an expert knowledge and experience of using plants.

Some professional garden designers are also landscape architects, a more formal level of training that usually requires an advanced degree and often a state license. Many amateur gardeners also attain a high level of experience from extensive hours working in their own gardens, through casual study or Master Gardener Programs offered by the American Horticultural Society.

LANDSCAPE PLAN

Landscape planning is a branch of landscape architecture. Urban park systems and greenways of the type planned by Frederick Law Olmsted are key examples of urban landscape planning. Landscape designers tend to work for clients who wish to commission construction work. Landscape planners can look beyond the 'closely drawn technical limits' and 'narrowly drawn territorial boundaries' which constrain design projects.

Landscape planners tend to work on projects which:

- Are of broad geographical scope
- Concern many land uses or many clients
- Are implemented over a long period of time

In rural areas, the damage caused by unplanned mineral extraction was one of the early reasons for a public demand for landscape planning.

A site plan is an architectural plan, and a detailed engineering drawing of proposed improvements to a given lot. A site plan usually shows a building footprint, travelways, parking, drainage facilities, sanitary sewer lines, water lines, trails, lighting, and landscaping. Such a plan of a site is a graphic representation of the arrangement of buildings, parking, drives, landscaping and any other structure that is part of a development project.

A site plan is a set of construction drawings that a builder or contractor uses to make improvements to a property. Counties can use the site plan to verify that development codes are being met and as a historical resource. Site plans are often prepared by a design consultant who must be either an licensed engineer, architect, landscape architect or land survey. The architect Map is part of a plan in Chandler, AZ.

A site plan is an architectural plan, and a detailed engineering drawing of proposed improvements to a given lot. A site plan usually shows a building footprint, travelways, parking, drainage facilities, sanitary sewer lines, water lines, trails, lighting, and landscaping. Such a plan of a site is a graphic representation of the arrangement of buildings, parking, drives, landscaping and any other structure that is part of a development project.

A site plan is a set of construction drawings that a builder or contractor uses to make improvements to a property. Counties can use the site plan to verify that development codes are being met and as a historical resource. Site plans are often prepared by a design consultant who must be either an licensed engineer, architect, landscape architect or land survey. The architect Map is part of a plan in Chandler, AZ.

Urban plan

Urban, city, and town planning is the integration of the disciplines of land use planning and transport planning, to explore a very wide range of aspects of the built and social environments of urbanized municipalities and communities. Regional planning deals with a still larger environment, at a less detailed level.

Based upon the origins of urban planning from the Roman (pre-Dark Ages) era, the current discipline revisits the synergy of the disciplines of urban planning, architecture and landscape architecture, varying upon from the interlectural strategic positioning from university to university.

Another key role of urban planning is urban renewal, and re-generation of inner cities by adapting urban planning methods to existing cities suffering from long-term infrastructural decay. The picture below is an architecture map of a part of south Chandler, AZ.

Architectural plan topics

Plan

A plan is typically any procedure used to achieve an objective. It is a set of intended actions, through which one expects to achieve a goal. Plans can be formal or informal:

- Structured and formal plans, used by multiple people, are more likely to occur in projects, diplomacy, careers, economic development, military campaigns, combat, or in the conduct of other business.
- Informal or ad-hoc plans are created by individuals in all of their pursuits.

The most popular ways to describe plans are by their breadth, time frame, and specificity; however, these planning classifications are not independent of one another. For instance, there is a close relationship between the short-and long-term categories and the strategic and operational categories. It is common for less formal plans to be created as abstract ideas, and remain in that form as they are maintained and put to use.

More formal plans as used for business and military purposes, while initially created with and as an abstract thought, are likely to be written down, drawn up or otherwise stored in a form that is accessible to multiple people across time and space. This allows more reliable collaboration in the execution of the plan

Plan view

A plan view is an orthographic projection of a 3-dimensional object from the position of a horizontal plane through the object. In other words, a plan is a section viewed from the top. In such views, the portion of the object in above the plane is omitted to reveal what lies beyond. In the case of a floor plan, the roof and upper portion of the walls may be omitted.

Roof plans are orthographic projections, but they are not sections as their viewing plane is outside of the object. A plan is a common method of depicting the internal arrangement of a 3-dimensional object in two dimensions. It is often used in technical drawing and is traditionally crosshatched. The style of crosshatching indicates the type of material the section passes through.

Planning

Planning in organizations and public policy is both the organizational process of creating and maintaining a plan; and the psychological process of thinking about the activities required to create a desired goal on some scale. As such, it is a fundamental property of intelligent behavior. This thought process is essential to the creation and refinement of a plan, or integration of it with other plans, that is, it combines forecasting of developments with the preparation of scenarios of how to react to them.

The term is also used to describe the formal procedures used in such an endeavor, such as the creation of documents diagrams, or meetings to discuss the important issues to be addressed, the objectives to be met, and the strategy to be followed. Beyond this, planning has a different meaning depending on the political or economic context in which it is used.

Technical Drawing

Technical drawing, also known as drafting, is the academic discipline of creating standardized technical drawings by architects, interior designers, drafters, design engineers, and related professionals. Standards and conventions for layout, line thickness, text size, symbols, view projections, descriptive geometry, dimensioning, and notation are used to create drawings that are ideally interpreted in only one way.

A person who does drafting is known as a drafter. In some areas this person may be referred to as a drafting technician, draftsperson, or draughtsman. This person creates technical drawings which are a form of specialized graphic communication. A technical drawing differs from a common drawing by how it is interpreted. A common drawing can hold many purposes and meanings, while a technical drawing is intended to concisely and clearly communicate all needed specifications to transform an idea into physical form.

Sketching

A sketch is a quickly executed freehand drawing that is not intended as a finished work. In general, a sketch is a quick way to record an idea for later use. Architect's sketches primarily serve as a way to try out different ideas and establish a composition before undertaking a more finished work, especially when the finished work is expensive and time consuming.

Architectural sketches, for example, are a kind of diagrams. These sketches, like metaphors, are used by architects as a mean of communication in aiding design collaboration. This tool helps architects to abstract attributes of hypothetical provisional design solutions and summarize their complex patterns, hereby enhancing the design process.

The basic drafting procedure is to place a piece of paper (or other material) on a smooth surface with right-angle corners and straight sides-typically a drawing board. A sliding straightedge known as a T-square is then placed on one of the sides, allowing it to be slid across the side of the table, and over the surface of the paper.

"Parallel lines" can be drawn simply by moving the T-square and running a pencil or technical pen along the T-square's edge, but more typically the T-square is used as a tool to hold other devices such as set squares or triangles. In this case the drafter places one or more triangles of known angles on the T-square-which is itself at right angles to the edge of the table-and can then draw lines at any chosen angle to others on the page. Modern drafting tables (which have by now largely been replaced by CAD workstations) come equipped with a drafting machine that is supported on both sides of the table to slide over a large piece of paper. Because it is secured on both sides, lines drawn along the edge are guaranteed to be parallel.

In addition, the drafter uses several tools to draw curves and circles. Primary among these are the compasses, used for drawing simple arcs and circles, and the French curve, typically a piece of plastic with complex curves on it. A spline is a rubber coated articulated metal that can be manually bent to most curves.

Drafting templates assist the drafter with creating recurring objects in a drawing without having to reproduce the object from scratch every time. This is especially useful when using common symbols; i.e. in the context of stagecraft, a lighting designer will typically draw from the USITT standard library of lighting fixture symbols to indicate the position of a common fixture across multiple positions. Templates are sold commercially by a number of vendors, usually customized to a specific task, but it is also not uncommon for a drafter to create their own templates.

This basic drafting system requires an accurate table and constant attention to the positioning of the tools. A common error is to allow the triangles to push the top of the T-square down slightly, thereby throwing off all angles. Even tasks as simple as drawing two angled lines meeting at a point require a number of moves of the T-square and triangles, and in general drafting can be a time consuming process.

A solution to these problems was the introduction of the mechanical "drafting machine", an application of the pantograph (sometimes referred to incorrectly as a "pentagraph" in these situations) which allowed the drafter to have an accurate right angle at any point on the page quite quickly. These machines often included the ability to change the angle, thereby removing the need for the triangles as well.

In addition to the mastery of the mechanics of drawing lines, arcs and circles (and text) onto a piece of paper-with respect to the detailing of physical objects-the drafting effort requires a thorough understanding of geometry, trigonometry and spatial comprehension, and in all cases demands precision and accuracy, and attention to detail of high order.

Although drafting is sometimes accomplished by a project engineer, architect-or even by shop personnel such as a machinist-skilled drafters (and/or designers) usually accomplish the task and are always in demand to some level.

Civil drawing

A civil drawing, or site drawing, is a type of technical drawing that shows information about grading, landscaping, or other site details. These drawings are intended to give a clear picture of all things in a construction site to a civil engineer.

Civil drafters work with civil engineers and other industry professionals to prepare models and drawings for civil engineering projects. Examples of civil engineering projects are bridges, building sites, canals, dams, harbors, roadways, railroads, pipelines, public utility systems, and waterworks. Civil drafters create maps, plans, cross sections, profiles, and detail drawings.

Process

The very early stages of a civil drawing start with surveying. Surveyors provide a map file of the job site from which civil designers and drafters develop drawings. After the surveying, other departments in the engineering firm start to work on other things such as draining, grading, foundation, and site preparation. Other projects such as electrical, piping, and sewage treatment might require another type of engineering or a specialized design. Final stages of a project may involve finished site preparation and landscape design. This aspect of a project may be contracted to landscape architects.

Standards

There are different sets of standards for each company, client, government, or professional organizations. For example The American Society of Mechanical Engineers (ASME) standards might not be the same standards that a client has. But there are also some standards that can be used in any field such as ASME Y14.2 Line Convections and Lettering.

Companies often times use standards that published by a professional organization but will add their own standards. Some examples of these standards can be having a different method of saving CADD data or plotting procedures. Government standards are city, county, or federal codes. These standards must always be followed even if the company or professional organization standards are available. Government standards apply even when the government is not the client. Project Standards are the final documents that are used when designing and building a project. These standards can include a combination of different standards from a company, client, or government. Before the drawings get released for bids or construction permitting they must get checked to see if they have all the appropriate standards, this is called standard checking. This process is usually done by a software.

Structural drawing

A structural drawing, a type of engineering drawing, is a plan or set of plans and details for how a building or other structure will be built. Structural drawings are generally prepared by registered professional engineers, and based on information provided by architectural drawings. The structural drawings are primarily concerned with the load-carrying members of a structure. They outline the size and types of materials to be used, as well as the general demands for connections. They do not address architectural details like surface finishes, partition walls, or mechanical systems. The structural drawings communicate the design of the building's structure to the building authority for review. Structural drawings are also included with a proposed building's contract documents, which guide contractors in detailing, fabricating, and installing parts of the structure.

The structural drawings set has different subsets: General Notes, Plans, Elevations, Sections, and Details

General Notes are part of structural drawings and they cover the codes used in design and the by-laws of the building. Typically there are no details on these drawings. Structural notes provide information regarding general material properties (steel or wood grade, concrete strength, etc) or construction requirements (soil compaction, weld procedures, etc). The structural notes also provide information about design criteria (gravity , seismic, and wind loading).

The structural plan drawings show the foundation, floor, and roof plan of the building. These plans provide information like size and location of the structural elements present in the respective plans.

Elevations show the exterior walls of a building or structure. In elevation drawings you can find the height of building (floors and roof elevations) and structural properties of elements present in the walls and that cannot be seen in plan drawings.

Sections plans are referenced in the plan view drawings and provide information about elements that cannot be see in plan drawings. The sections usually are cut through walls or structural elements that are not typical and the constructor needs to be aware off.

The details drawings provide particular information on how to construct or connect the structural elements. The details can be reference in plans, elevations and sections.

COMPUTER AIDED DESIGN

Today, the mechanics of the drafting task have largely been automated and accelerated through the use of computer-aided design systems (CAD). There are two types of computer-aided design systems used for the production of technical drawings" two dimensions ("2D") and three dimensions ("3D").

2D CAD systems such as AutoCAD replace the paper drawing discipline. The lines, circles, arcs and curves are created within the software. It is down to the technical drawing skill of the user to produce the drawing. There is still much scope for error in the drawing when producing first and third angle orthographic projections, auxiliary projections and cross sections. A 2D CAD system is merely an electronic drawing board. Its greatest strength over direct to paper technical drawing is in the making of revisions.

Where as in a conventional hand drawn technical drawing, if a mistake is found, or a modification is required, a new drawing must be made from scratch. The 2D CAD system allows a copy of the original to be modified, saving considerable time. 2D CAD systems can be used to create plans for large projects such as buildings and aircraft but provide no way to check the various components will fit together.

3D CAD systems such as Pro/ENGINEER first produce the geometry of the part, the technical drawing comes from user defined views of the part. Any orthographic, projected and section views are created by the software. There is no scope for error in the production of these views. The main scope for error comes in setting the parameter of first or third angle projection, and displaying the relevant symbol on the technical drawing. 3D CAD allows individual parts to be assembled together to represent the final product. Buildings, Aircraft, ships and cars are modeled, assembled and checked in 3D before technical drawings are released for manufacture.

Both 2D and 3D CAD systems can be used to produce technical drawings for any discipline. The various disciplines; electical, electronic, pneumatic, fluidic, etc., have industry recognised symbols to represent common components.

BS and ISO produce standards to show recommended practices but it is up to individuals to produce the drawings. There is no definitive standard for layout or style. The only standard across engineering workshop drawings is in the creation of orthographic projections and cross section views.

Drafting can represent two dimensions ("2D") and three dimensions ("3D") although the representation itself is always created in 2D (cf. Architectural model). Drafting is the integral communication of technical or engineering drawings and is the industrial arts sub-discipline that underlies all involved technical endeavors.

In representing complex, three-dimensional objects in two-dimensional drawings, the objects can be described by at least one view plus material thickness note, 2, 3 or as many views and sections that are required to show all features of object.

Applications for technical drawing

Architecture

The art and design that goes into making buildings is known as architecture. To communicate all aspects of the design, detailed drawings are used. In this field, the term plan is often used when referring to the full section view of these drawings. Architectural drawings describe and document an architect's design.

Engineering

Engineering can be a very broad term. It stems from the Latin ingenerare, meaning "to create". Because this could apply to everything that humans create, it is given a narrower definition in the context of technical drawing. Engineering drawings generally deal with mechanical engineered items, such as manufactured parts and equipment.

Engineering drawings are usually created in accordance with standardized conventions for layout, nomenclature, interpretation, appearance (such as typefaces and line styles), size, etc.

Its purpose is to accurately and unambiguously capture all the geometric features of a product or a component. The end goal of an engineering drawing is to convey all the required information that will allow a manufacturer to produce that component.

TECHNICAL ILLUSTRATIONS

Technical illustration is the use of illustration to visually communicate information of a technical nature. Technical illustrations can be component technical drawings or diagrams. The aim of technical illustration is "to generate expressive images that effectively convey certain information via the visual channel to the human observer".

The main purpose of technical illustration is to describe or explain these items to a more or less nontechnical audience. The visual image should be accurate in terms of dimensions and proportions, and should provide "an overall impression of what an object is or does, to enhance the viewer's interest and understanding".

According to Viola (2005) "illustrative techniques are often designed in a way that even a person with no technical understanding clearly understands the piece of art. The use of varying line widths to emphasize mass, proximity, and scale helped to make a simple line drawing more understandable to the lay person. Cross hatching, stippling, and other low abstraction techniques gave greater depth and dimension to the subject matter".

A cutaway drawing is a technical illustration, in which surface elements a three-dimensional model are selectively removed, to make internal features visible, but without sacrificing the outer context entirely.

The purpose of a cutaway drawing is to "allow the viewer to have a look into an otherwise solid opaque object. Instead of letting the inner object shine through the surrounding surface, parts of outside object are simply removed. This produces a visual appearance as if someone had cutout a piece of the object or sliced it into parts. Cutaway illustrations avoid ambiguities with respect to spatial ordering, provide a sharp contrast between foreground and background objects, and facilitate a good understanding of spatial ordering".

Technical drawings (the document)

Types of technical drawings

The two types of technical drawings are based on graphical projection. This is used to create an image of a three-dimensional object onto a two-dimensional surface.

Two-dimensional representation

Two-dimensional representation uses orthographic projection to create an image where only two of the three dimensions of the object are seen.

Three-dimensional representation

In three-dimensional representation, also referred to as pictorials, all three dimensions, of the three dimensions of an object, are visible.

Views

Multiview

Multiview is a type of orthographic projection. There are two conventions for using multiview, first-angle and third-angle. In both cases, the front or main side of the object is the same. First-angle is drawing the object sides based on where they land. Example, looking at the front side, rotate the object 90 degrees to the right. What is seen will be drawn to the right of the front side. Third-angle is drawing the object sides based on where they are. Example, looking at the front side, rotate the object 90 degrees to the right. What is seen is actually the left side of the object and will be drawn to the left of the front side

Section

While multiview relates to external surfaces of an object, section views show an imaginary plane cut through an object. This is often useful to show voids in an object.

Auxiliary

Auxiliary views utilize an additional projection plane other than the common planes in a multiview. Since the features of an object need to show the true shape and size of the object, the projection plane must be parallel to the object surface. Therefore, any surface that is not in line with the three major axis needs its own projection plane to show the features correctly.

Pattern

Patterns, sometimes called developments, show the size and shape of a flat piece of material needed for later bending or folding into a three dimensional shape.

Exploded

An exploded view drawing is a technical drawing of an object that shows the relationship or order of assembly of the various parts. It shows the components of an object slightly separated by distance, or suspended in surrounding space in the case of a three-dimensional exploded diagram. An object is represented as if there had been a small controlled explosion emanating from the middle of the object, causing the object's parts to be separated an equal distance away from their original locations.

An exploded view drawing (EVD) can show the intended assembly of mechanical or other parts. In mechanical systems usually the component closest to the center is assembled first, or is the main part in which the other parts get assembled. This drawing can also help to represent disassembly of parts, where the parts on the outside normally get removed first.

Patents

The applicant for a patent will be required by law to furnish a drawing of the invention whenever the nature of the case requires a drawing to understand the invention. This drawing must be filed with the application. This includes practically all inventions except compositions of matter or processes, but a drawing may also be useful in the case of many processes.

The drawing must show every feature of the invention specified in the claims, and is required by the patent office rules to be in a particular form. The Office specifies the size of the sheet on which the drawing is made, the type of paper, the margins, and other details relating to the making of the drawing. The reason for specifying the standards in detail is that the drawings are printed and published in a uniform style when the patent issues, and the drawings must also be such that they can be readily understood by persons using the patent descriptions.

SETS OF TECHNICAL DRAWINGS

Working drawings

Working drawings are the set of technical drawings used during the manufacturing phase of a product. In architecture, these typically include civil drawings, architectural drawings, structural drawings, mechanical systems drawings, electrical drawings, and plumbing drawings.

Assembly drawings

Assembly drawings show how different parts go together, identify those parts by number, and have a parts list, often referred to as a bill of materials.

Paper Size

Many paper size standards and conventions have existed at different times and in different countries. Today there is one widespread international ISO standard (including A4, B3, C4, etc.) and a localised standard used in North America (including letter, legal, ledger, etc.). The paper sizes affect writing paper, stationery, cards, and some printed documents. The standards also have related sizes for envelopes.

The international paper size standard, ISO 216, is based on the German DIN 476 standard for paper sizes. ISO paper sizes are all based on a single aspect ratio of square root of 2, or approximately 1:1.4142. The base A0 size of paper is defined to have an area of one m^2. With the given aspect ratio of square root of two, this corresponds to a piece of paper which its longer side is one metre multiplied by the square root of the square root (that is, the fourth root) of two and the shorter side being the inverse of this value. Rounded to millimetres the A0 paper size is 841 by 1,189 millimetres (33.1 × 46.8 in).

Successive paper sizes in the series A1, A2, A3, and so forth, are defined by halving the preceding paper size along the larger dimension. The most frequently used paper size is A4 (210 × 297 mm).

The main advantage of this system is its scaling: if a sheet with an aspect ratio of ?2 is divided into two equal halves parallel to its shortest sides, then the halves will again have an aspect ratio of ?2. Folded brochures of any size can be made by using sheets of the next larger size, e.g. A4 sheets are folded to make A5 brochures. The system allows scaling without compromising the aspect ratio from one size to another-as provided by office photocopiers, e.g. enlarging A4 to A3 or reducing A3 to A4. Similarly, two sheets of A4 can be scaled down and fit exactly 1 sheet without any cutoff or margins. Weights are easy to calculate as well: a standard A4 sheet made from 80 gram/m^2 paper weighs 5 grams (as it is one 16th of an A0 page, measuring 1 m^2), allowing one to easily compute the weight-and associated postage rate-by counting the number of sheets used.

The advantages of basing a paper size upon an aspect ratio of ?2 were already noted in 1786 by the German scientist and philosopher Georg Christoph Lichtenberg (in a letter to Johann Beckmann). Early in the twentieth century, Dr Walter Porstmann turned Lichtenberg's idea into a proper system of different paper sizes. Porstmann's system was introduced as a DIN standard (DIN 476) in Germany in 1922, replacing a vast variety of other paper formats. Even today the paper sizes are called "DIN A4" in everyday use in Germany. The term Lichtenberg ratio has recently been proposed for this paper aspect ratio.

In addition to the A series, there is a less common B series. The area of B series sheets is the geometric mean of successive A series sheets. So, B1 is between A0 and A1 in size, with an area of 0.707 m^2 (?0.5). As a result, B0 is 1 metre wide, and other sizes in the B series are a half, a quarter or further fractions of a metre wide. While less common in office use, it is used for a variety of special situations. Many posters use B-series paper or a close approximation, such as 50 cm×70 cm; B5 is a relatively common choice for books. The B series is also used for envelopes and passports.

The C series is used only for envelopes and is defined in ISO 269. The area of C series sheets is the geometric mean of the areas of the A and B series sheets of the same number; for instance, the area of a C4 sheet is the geometric mean of the areas of an A4 sheet and a B4 sheet. This means that C4 is slightly larger than A4, and B4 slightly larger than C4. The practical usage of this is that a letter written on A4 paper fits inside a C4 envelope, and a C4 envelope fits inside a B4 envelope.

The tolerances specified in the standard are

- ±1.5 mm (0.06 in) for dimensions up to 150 mm (5.9 in),
- ±2 mm (0.08 in) for lengths in the range 150 to 600 mm (5.9 to 23.6 in) and
- ±3 mm (0.12 in) for any dimension above 600 mm (23.6 in).

Tablet sizes

The sizes listed above are for paper sold loosely in reams. There are many sizes of tablets of paper, that is, sheets of paper bound at one edge, usually by a strip of plastic or hardened PVA adhesive. Often there is a pad of cardboard (also known as chipboard or greyboard) at the bottom of the stack. Such a tablet serves as a portable writing surface, and the sheets often have lines printed on them, usually in blue, to make writing in a line easier.

An older means of binding is to have the sheets stapled to the cardboard along the top of the tablet; there is a line of perforated holes across every page just below the top edge from which any page may be torn off. Lastly, a pad of sheets each weakly stuck with adhesive to the sheet below, trademarked as "Post-It" or "Stick-Em" and available in various sizes, serve as a sort of tablet.

"Letter pads" are 81?2 by 11 inches (215.9 by 279.4 mm), while the term "legal pad" is often used by laymen to refer to pads of various sizes including those of 81?2 by 14 inches (215.9 by 355.6 mm). There are "steno pads" (used by stenographers) of 6 by 9 inches (152.4 by 228.6 mm).

In countries where the ISO sizes are standard, most notebooks and tablets are sized to ISO specifications (for example, most newsagents in Australia stock A4 and A3 tablets).

Traditional inch-based paper sizes

Traditionally, a number of different sizes were defined for large sheets of paper, and paper sizes were defined by the sheet name and the number of times it had been folded. Thus a full sheet of "royal" paper was 25 × 20 inches, and "royal octavo" was this size folded three times, so as to make eight sheets, and was thus 10 by 6¼ inches.

Foolscap folio is often referred to simply as "folio" or "foolscap". Similarly, "quarto" is more correctly "copy draught quarto".

Many of these sizes were only used for making books, and would never have been offered for ordinary stationery purposes.

Demitab

The demitab or demi-tab (from the French "demi" or half tabloid) is 5.5 × 8.5 in (140 × 216 mm), equal to one quarter of a sheet of 11 × 17 in (279 × 432 mm) tabloid size paper. In actual circulation, the size 8 × 10.5 in (203 × 267 mm) is common for a demitab. Tabloid newspapers, which are "generally half the size of a broadsheet", also vary in size. To add to the lack of uniformity, broadsheets also vary in size.

Most industry standards express the direction of the grain last (e.g. 17×11 is short grain paper and 11×17 is long grain paper). See switching costs, network effects and standardization for possible reasons for differing regional adoption rates of the ISO standard sizes.

Transitional paper sizes

PA series

A transitional size called PA4 (210 × 280 mm/8.27 × 11.02 in) was proposed for inclusion into the ISO 216 standard in 1975. It has the height of Canadian P4 paper

(215 mm × 280 mm, about 8½ in × 11 in) and the width of international A4 paper (210 × 297 mm/8.27 × 11.69 in). The table to the right, shows how this format can be generalized into an entire format series.

The PA formats did not end up in ISO 216, because the committee felt that the set of standardized paper formats should be kept to the minimum necessary. However, PA4 remains of practical use today. In landscape orientation, it has the same 4:3 aspect ratio as the displays of traditional TV sets, some computer displays and data projectors. PA4, with appropriate margins, is therefore a good choice as the format of presentation slides. At the same time, PA4 is the largest format that fits on both A4 and US/Canadian Letter paper without resizing.

PA4 is used today by many international magazines, because it can be printed easily on equipment designed for either A4 or US Letter.

Antiquarian

Although the movement is towards the international standard metric paper sizes, on the way there from the traditional ones there has been at least one new size just a little larger than that used internationally. British architects and industrial designers once used a size called "Antiquarian" as listed above, but given in the New Metric Handbook (Tutt & Adler 1981) as 813 × 1,372 mm (32 × 54 in) for board size. This is a little larger than the A0 size. So for a short time, a size called A0a (1,000 × 1,370 mm/39.4 × 53.9 in) was used in Britain.

Art of Surveying

Surveying or land surveying is the technique and science of accurately determining the terrestrial or three-dimensional position of points and the distances and angles between them. These points are usually on the surface of the Earth, and they are often used to establish land maps and boundaries for ownership or governmental purposes.

To accomplish their objective, surveyors use elements of geometry, engineering, trigonometry, mathematics, physics, and law. An alternative definition, per the American Congress on Surveying and Mapping (ACSM), is the science and art of making all essential measurements to determine the relative position of points and/or physical and cultural details above, on, or beneath the surface of the Earth, and to depict them in a usable form, or to establish the position of points and/or details.

Furthermore, as alluded to above, a particular type of surveying known as "land surveying" (also per ACSM) is the detailed study or inspection, as by gathering information through observations, measurements in the field, questionnaires, or research of legal instruments, and data analysis in the support of planning, designing, and establishing of property boundaries.

It involves the re-establishment of cadastral surveys and land boundaries based on documents of record and historical evidence, as well as certifying surveys (as required by statute or local ordinance) of subdivision plats/maps, registered land surveys, judicial surveys, and space delineation. Land surveying can include associated services such as mapping and related data accumulation, construction layout surveys, precision measurements of length, angle, elevation, area, and volume, as well as horizontal and vertical control surveys, and the analysis and utilization of land survey data.

Surveying has been an essential element in the development of the human environment since the beginning of recorded history (about 5,000 years ago). It is required in the planning and execution of nearly every form of construction. Its most familiar modern uses are in the fields of transport, building and construction, communications, mapping, and the definition of legal boundaries for land ownership.

Types of surveys and applicability

- **ALTA/ACSM Survey:** a surveying standard jointly proposed by the American Land Title Association and the American Congress on Surveying and Mapping that incorporates elements of the boundary survey, mortgage survey, and topographic survey.
- **Archaeological survey:** used to accurately assess the relationship of archaeological sites in a landscape or to accurately record finds on an archaeological site.
- **As-built survey:** a survey carried out during or immediately after a construction project for record, completion evaluation and payment purposes. An as-built survey also known as a 'works as executed survey' documents the location of the recently constructed elements that are subject to completion evaluation. As built surveys are typically presented in red or redline and overlayed over existing design plans for direct comparison with design information.
- **Bathymetric survey:** a survey carried out to map the topography and features of the bed of an ocean, lake, river or other body of water.
- **Boundary survey:** a survey that establishes boundaries of a parcel using its legal description, which typically involves the setting or restoration of

monuments or markers at the corners or along the lines of the parcel, often in the form of iron rods, pipes, or concrete monuments in the ground, or nails set in concrete or asphalt.

- **Deformation survey:** a survey to determine if a structure or object is changing shape or moving. The three-dimensional positions of specific points on an object are determined, a period of time is allowed to pass, these positions are then re-measured and calculated, and a comparison between the two sets of positions is made.
- **Engineering surveys:** those surveys associated with the engineering design (topographic, layout and as-built) often requiring geodetic computations beyond normal civil engineering practise.
- **Foundation survey:** a survey done to collect the positional data on a foundation that has been poured and is cured. This is done to ensure that the foundation was constructed in the location, and at the elevation, authorized in the plot plan, site plan, or subdivision plan.
- **Geological survey:** generic term for a survey conducted for the purpose of recording the geologically significant features of the area under investigation. .
- **Hydrographic survey:** a survey conducted with the purpose of mapping the coastline and seabed for navigation, engineering, or resource management purposes.
- **Measured survey** : a building survey to produce plans of the building. such a survey may be conducted before renovation works, for commercial purpose, or at end of the construction process "as built survey"
- Mortgage survey or physical survey: a simple survey that delineates land boundaries and building locations. In many places a mortgage survey is required by lending institutions as a precondition for a mortgage loan.
- Soil survey, or soil mapping, is the process of determining the soil types or other properties of the soil cover over a landscape, and mapping them for others to understand and use.
- Structural survey: a detailed inspection to report upon the physical condition and structural stability of a building or other structure and to highlight any work needed to maintain it in good repair.
- Tape survey: this type of survey is the most basic and inexpensive type of land survey. Popular in the middle part of the 20th century, tape surveys

while being accurate for distance lack substantially in their accuracy of measuring angle and bearing. Standards that are practiced by professional land surveyors.

- Topographic survey: a survey that measures the elevation of points on a particular piece of land, and presents them as contour lines on a plot.

The basic principles of surveying have changed little over the ages, but the tools used by surveyors have evolved tremendously. Engineering, especially civil engineering, depends heavily on surveyors.

Whenever there are roads, railways, reservoir, dams, retaining walls, bridges or residential areas to be built, surveyors are involved. They establish the boundaries of legal descriptions and the boundaries of various lines of political divisions. They also provide advice and data for geographical information systems (GIS), computer databases that contain data on land features and boundaries.

Surveyors must have a thorough knowledge of algebra, basic calculus, geometry, and trigonometry. They must also know the laws that deal with surveys, property, and contracts.

In addition, they must be able to use delicate instruments with accuracy and precision. In the United States, surveyors and civil engineers use units of feet wherein a survey foot is broken down into 10ths and 100ths. Many deed descriptions requiring distance calls are often expressed using these units (125.25 ft). On the subject of accuracy, surveyors are often held to a standard of one one-hundredth of a foot; about 1/8th inch. Calculation and mapping tolerances are much smaller wherein achieving near-perfect closures are desired. Though tolerances such as this will vary from project to project, in the field and day to day usage beyond a 100th of a foot is often impractical.

In most of the United States, surveying is recognized as a distinct profession apart from engineering. Licensing requirements vary by state, but they generally have components of education, experience and examinations. In the past, experience gained through an apprenticeship, together with passing a series of state-administered examinations, was required to attain licensure. Now, most states insist upon basic qualification of a degree in surveying, plus experience and examination requirements.

The licensing process typically follows two phases. First, upon graduation, the candidate may be eligible to take the Fundamentals of Land Surveying exam, to be certified upon passing and meeting all other requirements as a surveyor in training (SIT). Upon being certified as an SIT, the candidate then needs to gain additional experience to become eligible for the second phase. That typically consists of the Principles and Practice of Land Surveying exam along with a state-specific examination.

Building surveying

Building surveying emerged in the 1970s as a profession in the United Kingdom by a group of technically minded General Practice Surveyors. Building surveying is a recognised profession in Britain, Australia and Hong Kong. In Australia in particular, due to risk mitigation and limitation factors, the employment of surveyors at all levels of the construction industry is widespread. There are still many countries where it is not widely recognized as a profession.

Services that building surveyors undertake are broad but can include:

- Construction design and building works
- Project management and monitoring
- Property Legislation advice
- Insurance assessment and claims assistance
- Defect investigation and maintenance advice
- Building surveys and measured surveys
- Handling planning applications
- Building inspection to ensure compliance with building regulations
- Pre-acquisition surveys
- Negotiating dilapidations claims

Building surveyors also advise on many aspects of construction including:

- Design
- Maintenance
- repair
- Refurbishment
- Restoration and preservation of buildings and monuments

Clients of a building surveyor can be the government agencies, businesses and individuals. Surveyors work closely with architects, planners, homeowners and tenants groups. Building surveyors may also be called to act as an expert witnesses. It is usual for building surveyors to earn a college degree before undertaking structured training to become a member of a professional organisation.

With the enlargement of the European community, the profession of the building surveyor is becoming more widely known in other European states, particularly France, where many English-speaking people buy second homes.

Lidar Surveying - Three-dimensional laser scanning provides high definition surveying for architectural, as-built, and engineering surveys. Recent technological advances make it the most cost-effective and time-sensitive solution for providing the highest level of detail available for interior and exterior building work.

The job of the land surveyor is to retrace legal description(s) from the deed belonging to the subject property by locating actual reference monumentation and verifying its correct position. Over time, development, vandalism and acts of nature often wreak havoc on monuments, so the land surveyor is often forced to consider other evidence such as fence locations, woodlines, monuments on neighboring property, parole evidence and other evidence.

SURVEYING TECHNIQUES

The simplest method for measuring height is with an altimeter - basically a barometer - using air pressure as an indication of height. But surveying requires greater precision. A variety of means, such as precise levels (also known as differential leveling), have been developed to do this. With precise leveling, a series of measurements between two points are taken using an instrument and a measuring rod. Differentials in height between the measurements are added and subtracted in a series to derive the net difference in elevation between the two endpoints of the series. With the advent of the Global Positioning System (GPS), elevation can also be derived with sophisticated satellite receivers, but usually with somewhat less accuracy than with traditional precise leveling. However, the accuracies may be similar if the traditional leveling would have to be run over a long distance.

Triangulation is another method of horizontal location made almost obsolete by GPS. With the triangulation method, distances, elevations and directions between objects at great distance from one another can be determined. Since the early days of surveying, this was the primary method of determining accurate positions of objects for topographic maps of large areas. A surveyor first needs to know the horizontal distance between two of the objects. Then the height, distances and angular position of other objects can be derived, as long as they are visible from one of the original objects. High-accuracy transits or theodolites were used for this work, and angles between objects were measured repeatedly for increased accuracy.

Many properties have considerable problems with regards to improper bounding, miscalculations in past surveys, titles, easements, and wildlife crossings.

Also many properties are created from multiple divisions of a larger piece over the course of years, and with every additional division the risk of miscalculation increases. The result can be abutting properties not coinciding with adjacent parcels, resulting in hiatuses (gaps) and overlaps. The art plays a role when a surveyor must solve a puzzle using pieces that do not exactly fit together. In these cases, the solution is based upon the surveyor's research and interpretation, along with established procedures for resolving discrepancies.

A land surveyor is an investigator of evidence. The land surveyor creates evidence on and under the ground to reference/preserve/perpetuate existing evidence. It is not the position of a land surveyor to make legal determinations; instead, the surveyor provides evidence that can be ruled on by a judge in a court of law (regarding legal decisions as to boundary lines). Evidence found and set by the land surveyor can be filed of record to be used for decisions as to land boundaries. In most States, it is the first one to the court house with that evidence that wins a boundary dispute unless other evidence is found with senior importance to the contrary. In most circumstances, the survey marker is of the highest priority as evidence of the boundary, unless evidence exists to show it was moved.

References to nearby survey markers are important when determining the "preponderance of evidence" for use by a judge who can set the legal boundary of land. A surveyor's opinion is valuable as evidence for legal boundary decisions, by those in authority, to legally settle a boundary dispute. In most cases, it is wiser for both parties to obtain the evidence and settle the dispute with the help of a land surveyor, than to press a suit in court.

It has become more common for title companies to try forcing a surveyor to change the measured distances and bearings to match existing documentation. If the title company is invested in the closing and promoting a faster "close" to avoid the cost of record corrections, this is a conflict of interest. Title company employees may not understand the importance of a bearing base and measured boundary, based on points found and missing points set, and based on the best fit provided by the surveyor.

This has become more and more of a problem with the lack of common knowledge of the importance of land surveying evidence. The survey boundary based on survey field evidence, especially measured boundary markers, should overrule previous written documentation that does not include the description of the survey markers found by the land surveyor.

Many do not understand the true meaning of a "metes and bounds" boundary description. The "bounds" or physical location and relationship of the survey markers has priority over the "metes" or measurements in the recorded description of a boundary. For example, an old measurement of 420 yards at a bearing of 120 degrees does not take priority over the actual positions of the survey markers on both ends, unless a marker is missing and needs to be re-set using that information. Other evidence that will verify the position of the missing marker, based on the senior evidence nearby as first priority, is preferred in such a case,

Surveying equipment

As late as the 1990s, the basic tools used in planar surveying were a tape measure for determining shorter distances, a level to determine height or elevation differences, and a theodolite, set on a tripod, to measure angles (horizontal and vertical), combined with the process of triangulation. Starting from a position with known location and elevation, the distance and angles to the unknown point are measured.

A more modern instrument is a total station, which is a theodolite with an electronic distance measurement device (EDM). A total station can also be used for leveling when set to the horizontal plane. Since their introduction, total stations have made the technological shift from being optical-mechanical devices to being fully electronic with an onship computer and software as well as humans.

Modern top-of-the-line total stations no longer require a reflector or prism (used to return the light pulses used for distancing) to return distance measurements, are fully robotic, and can even e-mail point data to the office computer and connect to satellite positioning systems, such as a Global Positioning System (GPS). Though real-time kinematic GPS systems have increased the speed of surveying, they are still horizontally accurate to only about 20 mm and vertically accurate to about 30-40 mm.

Total stations are still used widely, along with other types of surveying instruments. However, GPS systems do not work well in areas with dense tree cover or constructions. One-person robotic-guided total stations allow surveyors to gather precise measurements without extra workers to look through and turn the telescope or record data. A faster but expensive way to measure large areas (not details, and no obstacles) is with a helicopter, equipped with a laser scanner, combined with a GPS to determine the position and elevation of the helicopter. To increase precision, beacons are placed on the ground (about 20 km (12 mi) apart). This method reaches precisions between 5-40 cm (depending on flight height).

2

PLANNING OF INDUSTRIAL STRUCTURES

The industrial sector covers a range of building functions and types, including distribution centres, warehouses and small industrial units. The sector is characterised by a common requirement for long span structural solutions, and this article therefore also considers supermarkets, which typically share this requirement. Industrial is a key sector for structural steelwork. Standard steel portal frame construction dominates the sector, but other steel solutions, such as trusses, long span beams and fabricated sections may also be used to achieve the required long spans. The efficiency and cost effectiveness of the standard steel portal frame in this sector is reflected in the latest Construction Markets survey commissioned by the BCSA and Steel for Life. It shows that in 2021, steel frames accounted for 98.6% of all industrial shed building construction in the UK compared with 0.3% for insitu concrete, 0.3% for precast concrete and 0.8% for load bearing masonry.

This article focuses on the cost factors and the cost planning of steel framed industrial buildings. It builds on the general guidance on the cost of structural steelwork, and cost planning through design stages for structural steel framed buildings by considering the typical design and construction characteristics of industrial buildings. It explores the importance of programme, the key cost drivers for the industrial sector and the principal factors to consider when using standard cost ranges to ensure that realistic frame costs are achieved.

This article also examines the issue of sustainability with respect to frame selection, which is a further key factor in the owner occupier industrial sector, both as a method to reduce costs in use and as a key corporate driver.

INDUSTRIAL BUILDINGS: TYPICAL CHARACTERISTICS

The industrial sector comprises buildings with a wide range of uses, from out of town retail buildings and supermarkets, distribution centres and warehouses to science parks and light industrial buildings.

While these buildings have different functions and processes occurring within the space, their common design drivers of long spans, clear internal space and high bays to provide highly flexible internal layouts and to maximise floor plate efficiency result in a small range of typical structural design solutions across the sector.

The most commonly adopted structural solution to address these requirements is a steel portal frame. Portal frames are relatively lightweight structures that provide large clear internal spans and flexible space. The frame comprises columns and rafters, with localised haunches at the eaves to minimise rafter section size and at the apex to accommodate the connection between the rafters. Unlike other frames, the building envelope provides restraint to the structural steelwork, resulting in a very efficient structure. They typically include metal cladding and roofing but can accommodate a range of cladding materials where planning or aesthetic considerations are important.

A steel portal frame can efficiently deliver large spans from 25m to 40m with standard spacings between frames of 6m to 10m but typically 8m. Where multi-bay portals are required, a valley beam is often provided to remove alternate internal columns in a hit-and-miss arrangement to minimise the impact of the structure on the internal space.

Steel portal frames can also be constructed to a range of building heights to provide the high eaves required for certain building functions. For example, distribution centres typically include overhead craneage and therefore require a clear internal height of 10m to 13m or more, while supermarkets and warehouses are more likely to require 4m to 6m clear height depending on the storage racking system used.

As well as being able to provide clear internal height, the flexible nature of the space provided by a portal frame can also allow ancillary spaces to be incorporated easily within the main structure. Both industrial buildings and supermarkets will require a range of ancillary spaces from back of house offices to welfare and storage areas to be incorporated alongside the primary building function and in high eaves buildings it is common for these to be situated on first floor mezzanine floor space to avoid using the more valuable open plan spaces.

In addition to these benefits, a steel portal frame delivers a number of other advantages in the industrial sector. In particular, the lightweight nature of the frame can reduce the size of the required foundations and therefore the extent of the associated excavation, substructure and ground risk. This can be particularly beneficial on previously developed or urban sites, where substructure costs are a significant proportion of overall building costs.

A further key requirement of the sector, particularly for retail buildings, is speed of delivery of the building. A steel portal frame can offer programme advantages over other frame materials through both a quicker frame erection time, as elements can be delivered to site prefabricated and tend to be similar or repetitive, and also as noted above, potentially through reduced construction periods for substructure due to lighter frame weights. This can result in quicker delivery as well as reduced finance and cost of preliminaries.

While services provision is not the same issue for industrial buildings as it is for other sectors, if required, services can be integrated by providing openings in the webs of the rafters or alternatively using cellular beams for the rafters. However, in most cases, there is sufficient space below the frame to accommodate the services without compromising the operation of the building. It is most common to see the cellular beam arrangement on out of town retail buildings speculatively built to appeal to a range of potential tenants or purchasers.

While steel portal frames are the typically adopted frame type for the majority of industrial buildings, other steel solutions are available where there are more particular requirements. For example, steel trusses can be utilised for large distribution centres that require long spans as they can span from 50m to 80m and provide for services distribution within their depth. Where very long spans are required, space frames can be adopted; these are three dimensional trusses that span in two directions, are lightweight and can span up to 100m.

For developments that incorporate retail space on the ground floor but commercial or residential space at upper floor levels – a configuration that is becoming more common in the sector as a consequence of planning policy – long span beams can be used to provide ranges of 12m to 18m and offer relatively open plan retail space at ground floor level but also retain a grid that is compatible with other functions above.

Key frame cost drivers for industrial buildings

Given the existence of a number of common characteristics and requirements across the sector, it is not surprising that many industrial buildings and supermarkets have similar frame costs. Typically the frame cost range (excluding cladding rails and purlins) for low eaves (4m to 8m high) single storey industrial units adopting a steel portal frame solution will be between £105 and £138 per m2 GIFA (BCIS location index 100), however for high eaves buildings (10m to 13m high) the cost range will be higher as they have a higher steel frame weight per m2 GIFA and a range of £128 to £164 per m2 GIFA is more appropriate, as reflected in the cost table below.

However, as with all building types, it is still necessary during cost planning to consider a number of characteristics that can vary from site to site and project to project but which may impact on the structural frame cost of any individual building.

Location, site constraints and site configuration

For all building types, location, site constraints and site configuration should always be considered as the specific site will have a direct impact on the proposed building, influencing both the achievable design and the costs of construction. The characteristics of a proposed site will vary between an urban location and an out of town industrial complex or business park. The standard cost ranges are based on an out of town, less constrained location, which is typical for the majority of industrial buildings, so adjustments will be required for tight or city centre sites, which are likely to attract additional logistics and cost of preliminaries.

Site configuration can also impact on the structural design through floor plate arrangement. Where the structural grid has to change across the building to account for site factors or adjacent buildings, the efficiencies of repetition across the frame may not be realised. More repetitive structures provide material cost and on site erection efficiencies, so factors that reduce the level of repetition need to be assessed during planning as they may result in additional costs above those captured in the standard frame cost ranges.

Building height

Building height is a particularly important cost driver for industrial buildings and should be a key consideration during early cost planning when estimates are likely to be based on a frame weight per m2 of floor area. While the gross internal floor area may be the same, the weight of the steel frame will vary between a low and high eaves building on a kg per m2 basis, resulting in different overall frame costs per m2 GIFA.

Typical structural steel frame weights for low eaves buildings (4m to 8m high) are around 30-40kg per m2 GIFA including fittings, but are usually around 40-

50kg per m2 for high eaves buildings (10m to 13m high). For high eaves buildings it is therefore important to seek guidance from the structural engineer as to the proposed frame weight for the building as it could be around 20% higher than for a low eaves single storey portal framed building. Furthermore, should the proposed building have a very high bay configuration, with clear heights of up to 20m, adjustments will need to be made to the high eaves typical cost range to account for the further increased steel frame weight.

ANCILLARY ACCOMMODATION

Building height is generally driven by the building function and this is also a key driver of the requirements for ancillary accommodation and the structure needed to provide it, which will also impact on the overall building frame weight. For high eaves buildings, the extent of the proposed upper floor areas should be considered as this can vary significantly between different buildings, with ancillary space potentially being provided across as many as three mezzanine levels. The frame costs for these buildings will therefore need to be looked at carefully on a building by building basis with adjustments likely to be required to standard cost ranges.

Fire protection

A further cost consideration for industrial buildings is the required level of fire protection for the structure. Typically, fire protection is only required in single storey buildings where it is needed to satisfy boundary conditions or where there is a need to access the roof (e.g. for plant access). However for buildings with upper floor levels, mezzanines or internal offices the fire strategy will need to be clarified with the design team during cost planning to ensure that the extent and method of protection required is captured.

Design features

At the early design stages it is also important to gain an understanding of any design features that may require variations to the standard steel portal frame. For example, the incorporation of northlights for architectural, planning or site orientation reasons can result in an increase to the frame cost by as much as 35% due to the additional steelwork required to form the more complex roof profile.

Structural frame types and sustainability

A developing trend in recent years within the industrial and supermarket sectors is for the decision of building materials to increasingly be considered from a sustainability and long term value perspective, as well as the more traditional drivers of buildability, cost and programme.

This has not only been driven by government legislation and key requirements of building operational carbon performance, as set out in Part L of the Building Regulations, but also by owner occupier clients within the sector who have a strong incentive to reduce costs in use across the life of the building and through a recognition from the supermarket sector that sustainability can be a point of differentiation in the market.

It is therefore becoming more typical during the early design stages for assessments of sustainability to be undertaken across the main building elements to ensure that the material choices aide the achievement of the sustainability requirements for the project, whether driven by Part L or corporate goals. As part of the review of key building elements, the structural frame material choice is likely to be reviewed and it is therefore important to have an understanding of the relative sustainable performance of structural steel frames in the industrial sector.

Research on the relative sustainable performance of different structural frame types for both a distribution centre and supermarket has been carried out by AECOM and Sweett Group in the Target Zero project, which considered the performance of a number of frame types in terms of both operational carbon emissions and embodied energy.

Operational carbon

The Target Zero analysis demonstrated that in terms of operational carbon emissions, the frame material adopted does not have a significant impact on the overall performance of either the warehouse building (less than 3.5%) or supermarket (3.8%).

For the warehouse building, the report found that there was less than 1% variance in operational carbon emissions between a steel portal framed building and a precast concrete column and glulam beam structure, yet there was a 12% cost premium associated with the glulam and concrete option in terms of overall cost and an 83% premium on the cost of the structural frame itself.

Similarly, for the supermarket, it was also found that there was less than 1% variance in operational carbon emissions between the two considered frame options of a steel portal frame and a glulam structure, however there was a 2.4% cost premium for the glulam option overall driven by a 20-25% difference in the cost of the structural frame.

This research suggests that while some recent supermarket developments have incorporated a glulam or part glulam structure for sustainability reasons, despite the potential cost premium, the frame material choice actually only makes a very small difference to the overall operational carbon emissions of the building and the difference between the performance of the steel and glulam frames was less than 1% for both building types. Therefore, varying the frame material is unlikely to be a particularly cost effective way of achieving sustainable targets for proposed industrial or supermarket developments.

Indeed, the Target Zero studies demonstrated that for both the distribution warehouse building and the supermarket, consideration of the building lighting strategy was the most important factor in delivering cost-effective carbon savings, as lighting is typically the greatest energy demand in this sector. Indeed, the use of more efficient lighting systems alone delivered the 2010 Part L compliance target of reducing regulated carbon emissions by 25% for the distribution warehouse.

During the design stages, it is becoming more common to undertake assessments on the relative sustainable performance of different structural frame options by considering embodied carbon impacts alongside operational carbon emissions.

Embodied carbon

Embodied carbon assesses the total carbon dioxide emissions associated with the building but excluding the operational carbon occurring during the building use.

While Part L of the Building Regulations sets out assessment methodology and compliance requirements for operational carbon , consideration of embodied carbon is currently omitted, which can result in assessments following different methodologies.

However, it is important that assessments are undertaken firstly on a whole building level rather than on an elemental basis and secondly on a whole lifecycle (or cradle to cradle) basis.

Consideration of the whole building ensures that the intensity of material usage and the consequential impacts of frame choice on substructure are considered. It can often be overlooked that a significant proportion of a building's embodied carbon is in the substructure and that different framing materials have different implications for their design and embodied carbon impact, with lighter frames typically reducing the extent of foundations and excavation.

In addition, the whole building lifecycle should be assessed to determine the total embodied carbon impact. Some types of assessment omit the construction and end of life impacts and effectively assume that all materials perform in the same way. This is an oversimplification as there are differences in the end of life outcomes of materials, for example concrete is recovered and crushed for aggregate, steel and other metals are reused or recycled and a high proportion of timber goes to landfill.

The Target Zero reports demonstrated that while operational carbon emissions were similar between the different frame types, there were significant differences between the embodied energy performances for both the distribution warehouse and supermarket.

The adoption of the precast concrete and glulam structure for the distribution warehouse would have resulted in a 14% higher embodied carbon impact than the use of a steel portal framed structure and the glulam framed supermarket had a 2.4% higher embodied carbon impact than the steel portal frame .

This difference is largely due to differences in the end of life outcomes of steel and timber. Structural steel has high recycling and reuse rates compared with those for timber . As the timber in landfill decomposes it releases carbon dioxide and methane and while modern landfill sites are designed to collect methane, gas capture rates are typically 51% of all gas generated.

FACTORS TO BE CONSIDERED IN PLANNING FACTORY BUILDING

Factory building should be designed only after a complete production plan, plant layout and equipment sequences are determined so that the building exactly fits the production needs of the plant. It is necessary to keep in mind all factors that may affect the functioning of the plant in the building.

1. Nature of Manufacturing Process

The type of manufacturing process is the main determinant of plant buildings. The floor load, head space, bay size etc., depend on the type of machines and equipment to be used.

2. Plant Layout

The arrangement of machines, service centres and office exercise considerable influence on the design and construction of plant buildings. In fact, labour pattern should be determined first and the building should be just a shell around this design. However, provision for flexibility should be made to meet the future needs.

3. Space Requirements

The size of plant buildings depends upon space requirements for the flow and storage of materials, for location of machines, for service centres and for movement of employees. The height of the ceiling depends upon the type of equipment used. Use of overhead conveyors and tall equipment may require high roofs. But special structures and additional costs may be involved. Pits may be dug, if possible, to accommodate all equipment.

4. Material Handling

Ease in material handling assists in the reduction of manufacturing cycle time, avoids production bottlenecks and reduces material handling cost. Cranes, conveyor belts hoists, etc., are increasingly used for easy handling of materials. A reduction in the number of columns and the maintenance of the ceiling at a desirable height are significant to the use of material handling equipment. It is evident that the requirements of efficient material handling equipment affect the building design. Conversely, the characteristics of the factory building affect a firm's ability to use this equipment efficiently.

5. Plant Protection

The building should be so designed that there is adequate protection of plant from fire, theft, etc. Sprinkler system, fire escape exits, automatic alarms, outdoor hydrant, safety lights etc., may be used for this purpose.

6. Lighting

Lighting and illumination system within the plant exerts a significant influence on employees' productivity and fatigue. Therefore, lighting standards should be kept in view during plant planning.

7. Heating, Ventilation and Air Conditioning

It is quite common in large plants to provide separate structures for accommodating high pressure boilers. Their size will depend on heating requirements for production. Proper ventilation is necessary to provide adequate fresh air. Some manufacturing processes require considerable ventilation or controlled temperature and humidity. Air conditioning is particularly common in factory and other offices. Increasing costs of energy have led to new trends in heating, lighting and air conditioning. The statutory requirements of the Factories Act should also be met.

8. Service Facilities

Facilities relating to cooling towers, emergency power, compressed air, sewage treatment, etc., should also be considered in plant building. Waste disposal should be such that regulations regarding air and water pollution, etc., are not violated.

9. Accessibility

The factory building should be designed to ensure free movement of workers in the plant.

10. Aesthetic Considerations

Efforts should be made to make the building a pleasant place to work for the employees. Their comfort should receive top priority as they spend a great deal of their total working hours in the factory. The building should have an elegant appearance as this adds to the pride and prestige of the employees and the management. Any attractive, well designed plant promotes community goodwill. It also has an influence on employees' morale. Such a plant projects the progressive outlook of the organization. It is heartening to note that some managements have paid special attention to this factor at present requirement.

11. Appearance

The architectural style and building materials should be designed to give an attractive exterior to plant buildings. A pleasing appearance, good landscaping and clean surroundings are important to the local community.

12. Future Expansion

Future expansion needs should be considered in planning a factory building. If multi-storied buildings are to be expanded with additional floors, it must be planned in advance so that the original structure has sufficient footing to bear the additional weight.

13. Fire Protection

Need and importance for a fire protection system should be taken into account while planning for a factory building. This has become mandatory at present. Fire protection systems vary from the conventional fire extinguishers to automatic fire detectors and fire protection devices in industries.

14. Environmental Protection

Need to protect environment has to be considered while planning a plant layout. This is being vigorously advocated all over the world now. It has become statutory in India too. This includes ensuring greenbelt all around the factory, horticulture and effluent disposal and water/air treatment plant and wastage disposal schemes, etc.

15. Effluent disposal

Wherever chemical processes are used like electroplating, tanneries, etc., effluent treatment of the discharged water has become mandatory. Hence a factory building has to be planned for an effluent disposal system.

16. Air supply

While designing plan for a factory building, necessary care has to be taken for adequate Air supply. This can be obtained through adequate air compressors.

17. Contractors, consultants and collaborators

It is desirable that the building work is entrusted to well known consultants, experienced in constructing industrial buildings. The design of the factory and its layout also depends on the recommendation of collaborators — Indian and foreign when projects are executed as per technical collaboration from such parties.

18. Cost-economics

This is the most important aspect of a building. Better management planning and effective monitoring by use of modern managerial aids like PERT/Milestones can bring down the cost of construction considerably. Normally, the cost of construction goes up due to delay in execution, changes in drawing, interruption of work during construction, deletions of the original design.

Another source of wasteful expenditure is due to poor planning and procurement of materials, use of inferior quality materials and lack of effective supervision. So a factory building plan has to take into account of all cost factors.

3

IMPORTANCE OF BUILDING BYE-LAWS

Building bye-laws are a set of rules under which construction of a building needs to take place. The rules regulate coverage, height, architectural design and safety measures in order to protect buildings against natural disasters such as earthquakes and hazards such as fire as well as structural failures.

According to the definition, building bye laws are the norms set forth by the government authorities such as Ministry of Urban Development Affairs (MUDA), City Corporation and Developing Authorities, for instance, Bangalore Development Authority (BDA) in Bangalore. These norms are legal tools that regulate architectural and construction aspects of buildings to achieve orderly development in an area. They are crucial to protect buildings against fire, earthquake and structural failures. The development authority does not approve a building plan which fails to adhere to the bye laws.

The Town and Country Planning Organisation Ministry of Urban Development have devised a circular called "Model Building Bye-Laws - 2016" which states the regulatory mechanism and engineering parameters to keep in mind before starting any construction project in India. Many states in India, experience unauthorised constructions and building hazards due to absence of bye-laws. While states can form their own building bye-laws, Model Building Bye-Laws - 2016 acts as a standard for construction and a direction for states to devise their own rules for construction.

The first version of model building by-laws was prepared in 2003 by the Ministry of Urban Development when they incorporated provisions of structural safety, fire safety and barrier-free public buildings. The Bye-Laws were circulated to all state governments and union territories and out of 36 states and UTs, 22 have taken comprehensive revision in their respective building bye-laws since 2004.

The bye-laws were later revised in 2015 in light of emerging issues such as sanitation facilities for visitors in public areas, conservation of heritage buildings and barrier-free environment for disabled, elderly and children.

What new features does "Model Building Bye Laws-2016" include?

The circular issued in 2016, includes the following new provisions.

Green Building and Sustainable provisions - Rainwater harvesting, wastewater reuse, installation of solar rooftop PV rooms.

Provisions for disabled, elderly and children - Access path/ walkway, toilet, parking, drinking water, stairways, lifts, refuge and designing for children safety.

Structural safety and security - Includes disaster management guidelines for different building types. Prevention measure against 'soft storeys' in multi-storey buildings and proof checking of structural designs.

High rise building regulation - Structural safety requirements for high rise buildings, open spaces, building components, services and fire safety parameters.

Swachh Bharat Mission - Revised regulations for adequate toilet facilities for women and public conveniences such as separate public toilets for visitors in public buildings.

Rain water Harvesting - Provision for harvesting rainwater in different types of buildings and responsibilities for Urban Local Bodies (ULB) in public spaces. Includes rainwater harvesting techniques, provision for open spaces, enforcement and monitoring.

Communication technology - Provision for effects of electromagnetic radiations in built spaces by identifying emissions and sources.

Conservation of heritage sites - States the responsibilities for owners of heritage buildings regarding repairs, restrictions on development, preservation of the beauty of the area and penalties.

Special requirements for occupancy/land development - Includes provisions for industrial and commercial buildings regarding the development of land, building in the mining area, poultry farms and penal action for violation of the master plan.

Ease of doing business - Introduce provision for online building plan approval process, generation of reports/approvals, automatic generation of the completion certificate, clearance of master plans and integration of a "Single Window" process for users.

Building plans approval - Includes different bodies in charge for clearance of building plans before construction such as Urban Art Commission (UAC), National Monuments Authority (NMA), Airports Authority of India (AAI), Metro Rail Corporation (MRC), Heritage Conservation Committee (HCC), etc. to grant No Objection Certificate (NOC)/ approvals on the proposed building plans to the local bodies. Moreover, the clearance of the master plan from different bodies is required.

BUILDING BYE-LAWS' PURPOSE

Formulated primarily by a central authority, building bye-laws ensure that constructions are not only safe but also adhere to aesthetic standards. In that sense, these regulate the construction and the architectural aspects of construction activities. For example, the rules prescribed under the building bye-laws can make it mandatory for builders to keep fire safety and earthquake-resistance provisions at their projects. Building bye-laws also govern the provisions for open spaces in a project, with the aim to ensure that developments do not turn the city into a concrete jungle.

Building bye-laws also contain rules to ensure that there is minimum harm to the environment, as a result of developments.

As construction activities involve a lot of aspects that might be harmful or disturbing for those living in the surrounding areas, checks are also put in place, to keep such disturbances to their lowest level. Harmful levels of dust accumulation, health hazards, structural failure, risk of fire and high level of noise, are some of the aspects that builders need to take care of, throughout the construction cycle.

Machines used at the site can also emit radiations that are harmful to our health and the environment. To minimise the impact, buildings have to make efforts to mitigate electro-magnetic radiations in built spaces, by identifying appliances and sources of such emissions.

Hight of rooms and Buildings:

As per NBC, the general criteria to determine the height of a building is 1.5 times the width of the street to which the building abuts on its front side.

For street width of 8-12 m, the building height should not be more than 12 m

For street width more than 12 m, the height of a building should not be more than the width of the street and in case more than 24 m.

Lighting and ventilation of Rooms:

For sufficient lighting and ventilation of rooms of buildings, opening like windows and ventilators or direct opening should be provided.

The area of such openings excluding the area of doors should be minimum 1/10th of the floor area for dry and ho climate and minimum 1/6th of the area for wet hot climate.

The aggregate area of doors and windows should not be less than 1/7th of the floor area of the room.

Note: Foor area means built-up area excluding areas of walls.

Water supply and sanitary positions

Certain minimum water supply and sanitation convenience like water taps, sink, water closets washbasins etc. shall be provided as per NBC for different types of buildings.

Structural Designs (Size and section):

Each structure should be designed for safe loads, earthquake resistance, bearing capacities etc. as per relevant IS codes and NBC.

Some general thumb rule for structural design:

Depth of foundation: 0.75 m to 1.0 m for single-storeyed building below ground level and 1.0 m to 1.3 m for two-storeyed building.

Width of the foundation of the wall: Double the thickness of the wall just above the plinth and then add 30 cm to it will give the width of the foundation.

Concrete in Foundation of the wall: It should be nearly equal to 5/6th the thickness of wall above the plinth.

TYPES

Residential

Residential buildings are called houses/homes, though buildings containing large numbers of separate dwelling units are often called apartment buildings or apartment blocks to differentiate them from the more ‘individual' house.

Building types may range from one-room wood-framed, masonry, or adobe dwellings to multi-million dollar high-rise buildings able to house thousands of people. Increasing settlement density in buildings (and closer distances between buildings) is usually a response to high ground prices resulting from many people wanting to live close to work or similar attractors.

Multi-storey

A multi-storey building is a building that has multiple floors above ground in the building. Multi-story buildings aim to increase the area of the building without increasing the area of the land the building is built on, hence saving land and, in most cases, money (depending on material used and land prices in the area).

Creation

The practice of designing, constructing, and operating buildings is most usually a collective effort of different groups of professionals and trades. Depending on the size, complexity, and purpose of a particular building project, the project team may include:

- A real estate developer who secures funding for the project;
- One or more financial institutions or other investors that provide the funding
- Local planning and code authorities
- A Surveyor who performs an ALTA/ACSM and construction surveys throughout the project;
- Construction managers who coordinate the effort of different groups of project participants;
- Licensed architects and engineers who provide building design and prepare construction documents;
- Landscape architects;
- Interior designers;
- Other consultants;
- Contractors who provide construction services and install building systems such as climate control, electrical, plumbing, Decoration, fire protection, security and telecommunications;
- Marketing or leasing agents;
- Facility managers who are responsible for operating the building.

Regardless of their size or intended use, all buildings in the US must comply with zoning ordinances, building codes and other regulations such as fire codes, life safety codes and related standards.

Vehicles-such as trailers, caravans, ships and passenger aircraft-are treated as "buildings" for life safety purposes.

Ownership and funding

- Mortgage loan
- Real estate developer

Planning and design

- Architecture
- Building construction
- Civil engineering
- Mechanical, electrical, and plumbing design
- Quantity surveying
- Structural engineering
- Urban planning

Building services

Physical plant

Any building requires a certain amount of internal infrastructure to function, which includes such elements like heating / cooling, power and telecommunications, water and wastewater etc. Especially in commercial buildings (such as offices or factories), these can be extremely intricate sy.stems taking up large amounts of space (sometimes located in separate areas or double floors / false ceilings) and constitute a big part of the regular maintenance required.

Conveying systems

Systems for transport of people within buildings:

- Elevator
- Escalator
- Moving sidewalk (horizontal and inclined)

Systems for transport of people between interconnected buildings:

- Skyway
- Underground city

Building damage

Buildings may be damaged during the construction of the building or during maintenance. There are several other reasons behind building damage like accident. Buildings also may suffer from fire damage and flooding in special circumstances.

Building Code

A building code, or building control, is a set of rules that specify the minimum acceptable level of safety for constructed objects such as buildings and nonbuilding structures. The main purpose of building codes are to protect public health, safety and general welfare as they relate to the construction and occupancy of buildings and structures. The building code becomes law of a particular jurisdiction when formally enacted by the appropriate authority.

Building codes are generally intended to be applied by architects and engineers although this is not the case in the UK where Building Control Surveyors act as verifiers both in the public and private sector (Approved Inspectors), but are also used for various purposes by safety inspectors, environmental scientists, real estate developers, contractors and subcontractors, manufacturers of building products and materials, insurance companies, facility managers, tenants, and others.

There are often additional codes or sections of the same building code that have more specific requirements that apply to dwellings and special construction objects such as canopies, signs, pedestrian walkways, parking lots, and radio and television antennas.

PRACTICE OF BUILDING CODES

The practice of developing, approving, and enforcing building codes varies considerably among nations. In some countries building codes are developed by the government agencies or quasi-governmental standards organizations and then enforced across the country by the central government. Such codes are known as the national building codes (in a sense they enjoy a mandatory nation-wide application).

In other countries, where the power of regulating construction and fire safety is vested in local authorities, a system of model building codes is used. Model building codes have no legal status unless adopted or adapted by an authority having jurisdiction. The developers of model codes urge public authorities to reference model codes in their laws, ordinances, regulations, and administrative orders.

When referenced in any of these legal instruments, a particular model code becomes law. This practice is known as adoption by reference. When an adopting authority decides to delete, add, or revise any portions of the model code adopted, it is usually required by the model code developer to follow a formal adoption procedure in which those modifications can be documented for legal purposes.

There are instances when some local jurisdictions choose to develop their own building codes. At some point in time all major cities in the United States had their own building codes. However due to ever increasing complexity and cost of developing building regulations, virtually all municipalities in the country have chosen to adopt model codes instead.

For example, in 2008 New York City abandoned its proprietary 1968 New York City Building Code in favor of a customized version of the International Building Code. The City of Chicago remains the only municipality in America that continues to use a building code the city developed on its own as part of the Municipal Code of Chicago.

Similarly, in India, each municipality and urban development authority has its own building code, which is mandatory for all construction within their jurisdiction. All these local building codes are variants of a National Building Code, which serves as model code proving guidelines for regulating building construction activity.

Scope

Building codes generally include:

- Rules regarding parking and traffic impact
- Fire code rules to ensure safe evacuation in the event of a fire
- Requirements for earthquake, hurricane, tornado, flood, and tsunami resistance, especially in disaster prone areas or for very large buildings where a failure would be catastrophic
- Drainage, green space, and fence-building rules
- Requirements for specific building uses (for example, storage of flammable substances, or housing a large number of people)
- Energy provisions and consumption
- Grandfathering provisions: Unless the building is being renovated, the building code usually does not apply to existing buildings.
- Specifications on components
- Allowable installation methodologies
- Minimum and maximum room and exit sizes and location

- Qualification of individuals or corporations doing the work
- For high structures, anti-collision markers for the benefit of aircraft

Building codes are generally separate from zoning ordinances, but exterior restrictions (such as setbacks) may fall into either category.

PRESCRIPTIVE REQUIREMENTS

These requirements are usually a combination of prescriptive requirements that spell out exactly how something is to be done, and performance requirements which just outline what the required level of performance is and leave it up to the designer how this is achieved. Historically they are very reactive in that when a problem occurs the building codes change to ensure that the problem never happens again. In recent years there has been a move amongst most of the building codes to move to more performance requirements and less prescriptive requirements.

Traditionally building codes were generally short non complex interrelated sets of rules. They generally included reference to hundreds of other codes, standards and guidelines that specify the details of the component or system design, specify testing requirements for components, or outline good engineering practice.

These detailed codes required a great deal of specialization to interpret, and also greatly constrained change and innovation in building design. In recent years several countries, beginning with Australia, have moved to much shorter objective based buildings codes. Rather than prescribing specific details, objective codes lists a series of objectives all buildings must meet while leaving open how these objectives will be met. When applying for a building permit the designers must demonstrate how they meet each objective.

Antiquity

Building codes have a long history. What is generally accepted as the first building code was in the Code of Hammurabi which specified:

- 229. If a builder builds a house for someone, and does not construct it properly, and the house which he built falls in and kills its owner, then that builder shall be put to death.
- 230. If it kills the son of the owner, the son of that builder shall be put to death.
- 231. If it kills a slave of the owner, then he shall pay, slave for slave, to the owner of the house.

- 232. If it ruins goods, he shall make compensation for all that has been ruined, and inasmuch as he did not construct properly this house which he built and it fell, he shall re-erect the house from his own means.
- 233. If a builder builds a house for someone, even though he has not yet completed it; if then the walls seem toppling, the builder must make the walls solid from his own means.

The Law of Moses stipulated a specific construction requirement which is also an early form of a building code. The Bible book of Deuteronomy, chapter 22 verse 8, states:

- "In case you build a new house, you must also make a parapet for your roof, that you may not place bloodguilt upon your house because someone falling might fall from it."

Nineteenth-century building laws

The great changes in societies in Europe, the Industrial Revolution, the end of slavery in the United States, and immigration to the United States brought about the enactment of a number of building laws during the nineteenth century.

Baltimore passed its first building code in 1859. The Great Baltimore Fire occurred in February, 1904. Subsequent changes were made that matched other cities. In 1904, a Handbook of the Baltimore City Building Laws was published. It served as the building code for four years. Very soon, a formal building code was drafted and eventually adopted in 1908.

France

In Paris, great blocks of apartments were erected under the Second Empire (1852-70). The height of buildings was limited by law, so they were usually five or six stories at most.

Germany and Austria

Germany and Austria generally followed the French plan.

United Kingdom

The most important statutes of this kind in the United Kingdom were the London Building Act of 1844 and the Public Health Act of 1875. The Metropolitan Buildings Office was established in 1845.

United States

The major model building codes used in the United States are developed by the International Code Council (ICC), the International Building Code (IBC), the International Residential Code, the International Fire Code, the International Energy Conservation Code, the International Plumbing Code, the International Mechanical Code and others.

International Building Code

The International Building Code (IBC) is a model building code developed by the International Code Council (ICC). It has been adopted throughout most of the United States. Since the early 1900s, the system of building regulations in the United States was based on model building codes developed by three regional model code groups.

The codes developed by the Building Officials Code Administrators International (BOCA) were used on the East Coast and throughout the Midwest of the United States, while the codes from the Southern Building Code Congress International (SBCCI) were used in the Southeast and the codes published by the International Conference of Building Officials (ICBO) covered the West Coast and across to most of the Midwest.

Although regional code development has been effective and responsive to the regulatory needs of the local jurisdictions, by early 1990s it became obvious that the country needed a single coordinated set of national model building codes. The nation's three model code groups decided to combine their efforts and in 1994 formed the International Code Council (ICC) to develop codes that would have no regional limitations.

After three years of extensive research and development, the first edition of the International Building Code was published in 2000. The code was patterned on three legacy codes previously developed by the organizations that constitute ICC. By the year 2000, ICC had completed the International Codes series and ceased development of the legacy codes in favor of their national successor.

Legacy codes

- BOCA National Building Code (BOCA/NBC) by the Building Officials Code Administrators International (BOCA)
- Uniform Building Code (UBC) by the International Conference of Building Officials (ICBO)
- Standard Building Code (SBC) by the Southern Building Code Congress International (SBCCI)

COMPETING CODES AND FINAL ADOPTION

The National Fire Protection Association, initially, joined ICC in a collective effort to develop the International Fire Code (IFC). This effort however fell apart at the completion of the first draft of the document. Subsequent efforts by ICC and NFPA to reach agreement on this and other documents were unsuccessful, resulting in a series of disputes between the two organizations. After several failed attempts to find common ground with the ICC, NFPA withdrew from participation in development of the International Codes and joined with International Association of Plumbing and Mechanical Officials (IAPMO), American Society of Heating, Refrigerating and Air-Conditioning Engineers (ASHRAE) and the Western Fire Chiefs Association to create an alternative set of codes. First published in 2002, the code set named the Comprehensive Consensus Codes, or C3, includes the NFPA 5000 building code as its centerpiece and the companion codes such as the National Electrical Code, NFPA 101 Life Safety Code, UPC, UMC, and NFPA 1. Unlike the IBC, the NFPA 5000 conformed to ANSI-established policies and procedures for the development of voluntary consensus standards.

The NFPA's move to introduce a competing building standard received strong opposition from powerful trade groups such as the American Institute of Architects (AIA), BOMA International and National Association of Home Builders (NAHB). Subsequent to unsuccessful attempts to encourage peaceful cooperation and resolution between NFPA and ICC on their codes disputes, a number of organizations, including AIA, BOMA and two dozen commercial real estate associations, founded the Get It Together coalition, which repeatedly urged NFPA to abandon code development related to NFPA 5000 and to work with ICC to integrate the other NFPA codes and standards into the ICC family of codes.

Initially, California adopted the NFPA 5000 codes as a baseline for the future California Building Code, but later rescinded the decision when Gov. Davis-who had strong ties with the State Fire Marshall who, in turn, supported NFPA 5000-was recalled from office and Gov Schwarzenegger was elected. Upon his election, Gov. Schwarzenegger rescinded directive to use NFPA 5000, and California adopted the IBC. The main driver for this decision were increased costs involved in training architects and engineers to design for a new code, construction contractors to build to the new code, and state and local agencies to have to learn a new code with which to inspect. Additionally, NFPA 5000 would cause a disparity between California and the majority of other states which had adopted IBC; not to mention, the legacy ICBO started in California and was headquartered in Whittier, CA.

Since then, benefits of using one code for the entire country has been realized by all. At this time, jurisdictions within all 50 states and the District of Columbia have adopted the IBC.

A large portion of the International Building Code deals with fire prevention. It differs from the related International Fire Code in that the IBC addresses fire prevention in regard to construction and design and the fire code addresses fire prevention in regard to the operation of a completed and occupied building. For example, the building code sets criteria for the number, size and location of exits in the design of a building while the fire code requires the exits of a completed and occupied building to be unblocked. The building code also deals with access for the disabled and structural stability (including earthquakes). The International Building Code applies to all structures in areas where it is adopted, except for one and two family dwellings.

Parts of the code reference other codes including the International Plumbing Code, the International Mechanical Code, the National Electric Code, and various National Fire Protection Association standards. Therefore, if a municipality adopts the International Building Code, it also adopts those parts of other codes referenced by the IBC. Often, the plumbing, mechanical, and electric codes are adopted along with the building code.

The code book itself (2000 edition) totals over 700 pages and chapters include:

- Building occupancy classifications
- Building heights and areas
- Interior finishes
- Foundation, wall, and roof construction
- Fire protection systems (sprinkler system requirements and design)
- Materials used in construction
- Elevators and escalators
- Already existing structures
- Means of egress

Means of Egress

The phrase "means of egress" refers to the ability to exit the structure, primarily in the event of an emergency, such as a fire. Specifically, a means of egress is broken into three parts: the path of travel to an exit, the exit itself, and the exit discharge (the path to a safe area outside). The code also address the number of exits required for a structure based on its intended occupancy use and the number of people who could be in the place at one time as well as their relative locations. It also deals with special needs, such as hospitals, nursing homes, and prisons where evacuating people may have special requirements. In some instances, requirements are made based on possible hazards (such as in industries) where flammable or toxic chemicals will be in use.

"Accessibility" refers to the accommodation of physically challenged people in structures. This includes building entry, parking spaces, elevators, and restrooms.

Building code requirements generally apply to the construction of new buildings and alterations or additions to existing buildings, changes in the use of buildings, and the demolition of buildings or portions of buildings at the ends of their useful or economic lives. As such, building codes obtain their effect from the voluntary decisions of property owners to erect, alter, add to, or demolish a building in a jurisdiction where a building code applies .

Alterations and additions to an existing building must usually comply with all new requirements applicable to their scope as related to the intended use of the building. Some changes in the use of a building often expose the entire building to the requirement to comply fully with provisions of the code applicable to the new use. Some jurisdictions limit such application to matters of fire safety, disabled access or structural integrity, others apply an economic feasibility or practicality test, and still others exempt buildings of special use or architectural or historic significance.

Existing buildings are not exempt from new requirements, especially those considered essential to achieve health, safety or general welfare objectives of the adopting jurisdiction, even when they are not otherwise subject to alteration, addition, change in use, or demolition. Such requirements typically remedy existing conditions, considered in hindsight, inimical to safety, such as the lack of automatic fire sprinklers in certain places of assembly, as became a major concern after the Station nightclub fire in 2003 killed 100 people.

Although such remedial enactments address existing conditions, they do not violate the United States Constitution's ban on the adoption of ex post facto law, as they do not criminalize or seek to punish past conduct . Such requirements merely prohibit the maintenance or continuance of conditions that would prove injurious to a member of the public or the broader public interest.

Assertions by property rights advocates in the United States that such requirements violate the "takings clause" of the Fifth Amendment to the United

States Constitution, have generally failed on grounds that compliance with such requirements increases rather than decreases the capital value of the property concerned.

Some states especially those that delegate their adoption and enforcement authority to subordinate local jurisdictions, may exempt their own buildings from compliance with local building codes or local amendments to a statewide building code. Similarly, property owned by the United States Government is considered exempt from state and local enactments, although such properties are generally not exempt from inspection by state or local authorities, except on grounds of protecting national defense or national security.

In lieu of submitting themselves to compliance with the requirements of other government jurisdictions, most state and federal agencies adopt construction and maintenance requirements that either reference model building codes or model their provisions on their requirements.

Some jurisdictions have enacted requirements to bring certain types or uses of existing buildings into compliance with new requirements, such as the installation of smoke alarms in households or dwelling units, at the time of sale. Some safety advocates have suggested a similar approach to encourage remedial application of other requirements, but few jurisdictions have found it economical or equitable to disincentivise property transactions in this way .

Many jurisdictions have found the application of new requirements to old, particularly historic buildings, challenging. New Jersey, for example, has adopted specific state amendments to provide a means of code compliance to existing structures without forcing the owner to comply with rigid requirements of the currently adopted Building Codes where it may be technically infeasible to do so.

California has also enacted a specific historic building code. Other states require compliance with building and fire codes, subject to reservations, limitations, or jurisdictional discretion to protect historic building stock as a condition of nominating or listing a building for preservation or landmark status, especially where such status attracts tax credits, investment of public money, or other incentives.

The listing of a building on the National Register of Historic Places does not exempt it from compliance with state or local building code requirements.

Updating Cycle

Updated editions of the IBC are published on a three year cycle (2000, 2003, 2006...). This fixed schedule has led other organizations, which produce referenced standards, to align their publishing schedule with that of the IBC .

Referenced Standards

Model building codes rely heavily on referenced standards as published and promulgated by other standards organizations such as ASTM (ASTM International), ANSI (American National Standards Institute), and NFPA (National Fire Protection Association). The structural provisions rely heavily on referenced standards, such as the Minimum Design Loads for Buildings and Structures published by the American Society of Civil Engineers (ASCE-7).

Changes in parts of the reference standard can result in disconnection between the corresponding editions of the reference standards.

Codes

Most states or municipalities in the United States of America adopt the ICC family of codes and other reference standards and codes in whole or with local amendments. This has the effect of designating a copyrighted work as actual law , and, once enacted, that law enters the public domain . The model codes themselves, prior to their adoption as law, are not in the public domain and the ICC retains copyright on the model code itself.

ICC Building codes

- International Building Code
- International Residential Code
- International Fire Code
- International Plumbing Code
- International Mechanical Code
- International Fuel Gas Code
- International Energy Conservation Code
- ICC Performance Code
- International Wildland Urban Interface Code
- International Existing Building Code
- International Property Maintenance Code

Fire Prevention

Fire safety refers to precautions that are taken to prevent or reduce the likelihood of a fire that may result in death, injury, or property damage, alert those in a structure to the presence of a fire in the event one occurs, better enable those threatened by a fire to survive, or to reduce the damage caused by a fire. Fire safety measures include those that are planned during the construction of a building or implemented in structures that are already standing, and those that are taught to occupants of the building.

Threats to fire safety are referred to as fire hazards. A fire hazard may include a situation that increases the likelihood a fire may start or may impede escape in the event a fire occurs.

Fire safety is often a component of building safety. Those who inspect buildings for violations of the Fire Code and go into schools to educate children on Fire Safety topics are fire department members known as fire prevention officers. The Chief Fire Prevention Officer or Chief of Fire Prevention will normally train newcomers to the Fire Prevention Division and may also conduct inspections or make presentations.

Of course, the best way to practice fire safety is to make sure a fire doesn't break out in the first place. That means you should always be aware of potential hazards in your home. Start by keeping these tips in mind.

Important tips:

- Replace or professionally repair any appliances that spark, smell unusual, or overheat.
- Don't run electrical wires under rugs.
- Make sure lamps and night-lights are not touching bedspreads, drapes, or other fabrics.
- Use caution when using electric blankets.
- Don't let kids use kitchen appliances by themselves and supervise any art or science projects that involve electrical devices.
- Cover any outlets that are not in use with plastic safety covers if you have toddlers or young children in your home.

Key elements of a fire safety policy

- Building a facility in accordance with the version of the local building code
- Maintaining a facility and conducting yourself in accordance with the provisions of the fire code. This is based on the occupants and operators of the building being aware of the applicable regulations and advice.

Examples of these include:

- Not exceeding the maximum occupancy within any part of the building.
- Maintaining proper fire exits and proper exit signage (e.g., exit signs pointing to them that can function in a power failure)
- Placing and maintaining fire extinguishers in easily accessible places.
- Properly storing/using, hazardous materials that may be needed inside the building for storage or operational requirements (such as solvents in spray booths).
- Prohibiting flammable materials in certain areas of the facility.
- Periodically inspecting buildings for violations, issuing Orders To Comply and, potentially, prosecuting or closing buildings that are not in compliance, until the deficiencies are corrected or condemning it in extreme cases.
- Maintaining fire alarm systems for detection and warning of fire.
- Obtaining and maintaining a complete inventory of firestops.
- Ensuring that spray fireproofing remains undamaged.
- Maintaining a high level of training and awareness of occupants and users of the building to avoid obvious mistakes, such as the propping open of fire doors.
- Conduct fire drills at regular intervals throughout the year.

Some common fire hazards are:

- Electrical systems that are overloaded, resulting in hot wiring or connections, or failed components
- Combustible storage areas with insufficient protection
- Combustibles near equipment that generates heat, flame, or sparks
- Candles
- Smoking (Cigarettes, cigars, pipes, etc.)
- Equipment that generates heat and utilizes combustible materials
- Flammable liquids
- Fireplace chimneys not properly or regularly cleaned
- Cooking appliances - stoves, ovens
- Heating appliances - wood burning stoves, furnaces, boilers, portable heaters
- Electrical wiring in poor condition

- Batteries
- Personal ignition sources - matches, lighters
- Electronic and electrical equipment
- Exterior cooking equipment - BBQ
- Campfires

FIRE CODE

In America, the Fire code (also Fire prevention code or Fire safety code) is a model code adopted by the state or local jurisdiction and enforced by fire prevention officers within municipal fire departments. It is a set of rules prescribing minimum requirements to prevent fire and explosion hazards arising from storage, handling, or use of dangerous materials, or from other specific hazardous conditions. It complements the building code.

The fire code is aimed primarily at preventing fires, ensuring that necessary training and equipment will be on hand, and that the original design basis of the building, including the basic plan set out by the architect, is not compromised. The fire code also addresses inspection and maintenance requirements of various fire protection equipment in order to maintain optimal active fire protection and passive fire protection measures.

A typical fire safety code includes administrative sections about the rule-making and enforcement process, and substantive sections dealing with fire suppression equipment, particular hazards such as containers and transportation for combustible materials, and specific rules for hazardous occupancies, industrial processes, and exhibitions.

Sections may establish the requirements for obtaining permits and specific precautions required to remain in compliance with a permit. For example, a fireworks exhibition may require an application to be filed by a licensed pyrotechnician, providing the information necessary for the issuing authority to determine whether safety requirements can be met. Once a permit is issued, the same authority (or another delegated authority) may inspect the site and monitor safety during the exhibition, with the power to halt operations, when unapproved practices are seen or when unforeseen hazards arise.

List of some typical fire and explosion issues in a fire code

- fireworks, explosives, mortars and cannons, model rockets (licenses for manufacture, storage, transportation, sale, use)

- Certification for servicing, placement, and inspecting fire extinguishing equipment
- General storage and handling of flammable liquids, solids, gases (tanks, personnel training, markings, equipment)
- Limitations on locations and quantities of flammables (e.g., 10 liters of gasoline inside a residential dwelling)
- Specific uses and specific flammables (e.g., dry cleaning, gasoline distribution, explosive dusts, pesticides, space heaters, plastics manufacturing)
- Permits and limitations in various building occupancies (assembly hall, hospital, school, theater, elderly care, child care, prs that require a smoke detector, sprinkler system, fire extinguisher, or other specific equipment or procedures
- Removal of interior and exterior obstructions to emergency exits or firefighters and removal of hazardous materials
- Permits and limitations in special outdoor applications (tents, asphalt kettles, bonfires, etc.)
- other hazards (flammable decorations, welding, smoking, bulk matches, tire yards)
- Electrical safety code

Fuel gas code

Public fire safety education

Fire prevention programs may include distribution of smoke detectors, visiting schools to review key topics with the students and implementing nationally recognized programs such as NFPAs "Risk Watch" & "Learn not to burn."

Other programs or props can be purchased by fire departments or community organizations. These are usually entertaining and designed to capture children's attention and relay important messages. Props include those that are mostly auditory, such as puppets & robots. The prop is visually stimulating but the safety message is only transmitted orally. Other props are more elaborate, access more senses and increase the learning factor. They mix audio messages and visual queues with hands-on interaction. Examples of these include mobile trailer safety houses and tabletop hazard house simulators.

All programs tend to mix messages of general injury prevention, safety, fire prevention and escape in case of fire. In most cases the fire department representative is regarded as the expert and is expected to present information in a manner that is appropriate for each age group.

ELEVATOR SAFETY

Elevators are generally powered by electric motors that either drive traction cables or counterweight systems like a hoist, or pump hydraulic fluid to raise a cylindrical piston like a jack. Because of wheelchair access laws, elevators are often a legal requirement in new multi-storey buildings, especially where wheelchair ramps would be impractical.

Pneumatic Vacuum Elevators operate without cables or pistons and can be installed more easily and quickly than their alternatives since their housing is composed of prefabricated sections which are considerably narrower than conventional lift shafts. These sections are transparent and afford the passenger a near 360° view. Other notable features of the vacuum elevator are as follows:

- No pit excavation, hoist way, or machine room needed.
- Installation within one to two days
- Two to four stops for residential, marine, and stage use (35ft. total rise)
- Ideal for new and existing homes due to the minimal space needed to fit the elevator
- Self-supporting structure: the elevator can rest on any existing ground floor
- "Green Elevator" : minimal energy consumption used during ascent and no energy used during descent
- Minimal maintenance: no oils or lubrication required for the elevator
- Absolute safety in case of a power failure since the moving car automatically descends to the lowest level and the electro-mechanical door will open to let the passenger out
- Elevator runs on 220Volts and cabin electric circuits are 24 volts, eliminating the risk of shock

Cable-borne elevators

Statistically speaking, elevators are extremely safe. Their safety record is unsurpassed by any other vehicle system. In 1998, it was estimated that approximately eight 100-millionths of one percent (1 in 12 million) of elevator rides resulted in an anomaly, and the vast majority of these were minor things such as the doors failing to open.

For all practical purposes, there are no cases of elevators simply free-falling and killing the passengers inside; of the 20 to 30 elevator-related deaths each year, most of them are maintenance-related - for example, technicians leaning too far into the shaft or getting caught between moving parts, and most of the rest are attributed to easily avoidable accidents, such as people stepping blindly through doors that open into empty shafts or being strangled by scarves caught in the doors.

In fact, prior to the September 11th terrorist attacks, the only known free-fall incident in a modern cable-borne elevator happened in 1945 when a B-25 bomber struck the Empire State Building in fog, severing the cables of an elevator cab, which fell from the 75th floor all the way to the bottom of the building, seriously injuring (though not killing) the sole occupant - the female elevator operator.

However, there was an incident in 2007 at a Seattle children's hospital, where a ThyssenKrupp ISIS machine room-less elevator free-fell until the safety brakes were engaged. This was due to a flaw in the design where the cables were connected at one common point, and the kevlar ropes had a tendency to overheat and cause slipping (or, in this case, a free-fall). While it is possible (though extraordinarily unlikely) for an elevator's cable to snap, all elevators in the modern era have been fitted with several safety devices which prevent the elevator from simply free-falling and crashing.

An elevator cab is typically borne by six or eight hoist cables, each of which is capable on its own of supporting the full load of the elevator plus twenty-five percent more weight. In addition, there is a device which detects whether the elevator is descending faster than its maximum designed speed; if this happens, the device causes copper brake shoes to clamp down along the vertical rails in the shaft, stopping the elevator quickly, but not so abruptly as to cause injury. This device is called the governor, and was invented by Elisha Graves Otis. In addition, a hydraulic buffer is installed at the bottom of the shaft to somewhat cushion any impact.

Hydraulic elevators

Past problems with early hydraulic elevators meant those built prior to a code change in 1972 were subject to possible catastrophic failure. The code had previously required only single-bottom hydraulic cylinders. In the event of a cylinder breach, an uncontrolled fall of the elevator might result.

Because it is impossible to verify the system completely without a pressurized casing (as described below), it is necessary to remove the piston to inspect it. The cost of removing the piston is such that it makes no economic sense to re-install the old cylinder; therefore it is necessary to replace the cylinder and install a new piston. Another solution to protect against a cylinder blowout is to install a "life jacket."

This is a device which, in the event of an excessive downward speed, clamps onto the cylinder and stops the car. A device known as a rupture valve is often attached to the hydraulic inlet/outlet of the piston and can be adjusted for a maximum flow rate. If a pipe or hose were to break (rupture), the flow rate of the rupture valve will surpass a set limit and mechanically stop the outlet flow of hydraulic fluid, thus stopping the piston and the car in the down direction.

In addition to the safety concerns for older hydraulic elevators, there is risk of leaking hydraulic oil into the aquifer and causing potential environmental contamination. This has led to the introduction of PVC liners (casings) around hydraulic cylinders which can be monitored for integrity.

In the past decade, recent innovations in inverted hydraulic jacks have eliminated the costly process of drilling the ground to install a borehole jack. This also eliminates the threat of corrosion to the system and increases safety.

Mine-shaft elevators

Safety testing of mine shaft elevator rails is routinely undertaken. The method involves destructive testing of a segment of the cable. The ends of the segment are frayed, then set in conical zinc molds. Each end of the segment is then secured in a large, hydraulic stretching machine. The segment is then placed under increasing load to the point of failure. Data about elasticity, load, and other factors is compiled and a report is produced. The report is then analyzed to determine whether or not the entire rail is safe to use.

A passenger elevator is designed to move people between a building's floors. Passenger elevators capacity is related to the available floor space. Generally passenger elevators are available in capacities from 1,000 to 6,000 pounds (450-2,700 kg) in 500 lb (230 kg) increments. Generally passenger elevators in buildings eight floors or less are hydraulic or electric, which can reach speeds up to 200 ft/min (1.0 m/s) hydraulic and up to 500 ft/min electric. In buildings up to ten floors, electric and gearless elevators are likely to have speeds up to 500 ft/min (2.5 m/s), and above ten floors speeds begin at 500 ft/min (2.5 m/s) up to 2000 ft/min (10 m/s).

Sometimes passenger elevators are used as a city transport along with funiculars. For example, there is a 3-station underground public elevator in Yalta, Ukraine, which takes passengers from the top of a hill above the Black Sea on which hotels are perched, to a tunnel located on the beach below. At Casco Viejo station in the Bilbao Metro, the elevator that provides access to the station from a hilltop neighborhood doubles as city transportation: the station's ticket barriers are set up in such a way that passengers can pay to reach the elevator from the entrance in the lower city, or vice versa. See also the Elevators for urban transport section. The Equitable Building completed in 1870 in New York City was the first office building to have passenger elevators.

TYPES OF PASSENGER ELEVATORS

Passenger elevators may be specialized for the service they perform, including: hospital emergency (Code blue), front and rear entrances, a television in high rise buildings, double decker, and other uses. Cars may be ornate in their interior appearance, may have audio visual advertising, and may be provided with specialized recorded voice instructions.

An express elevator does not serve all floors. For example, it moves between the ground floor and a skylobby, or it moves from the ground floor or a skylobby to a range of floors, skipping floors in between. These are especially popular in eastern Asia.

Capacity

Residential elevators may be small enough to only accommodate one person while some are large enough for more than a dozen. Wheelchair, or platform lifts, a specialized type of elevator designed to move a wheelchair 6 ft (1.8 m) or less, often can accommodate just one person in a wheelchair at a time with a load of 1000 lb (450 kg).

Freight elevators

A freight elevator, or goods lift, is an elevator designed to carry goods, rather than passengers. Freight elevators are generally required to display a written notice in the car that the use by passengers is prohibited (though not necessarily illegal), though certain freight elevators allow dual use through the use of an inconspicuous riser.

Freight elevators are typically larger and capable of carrying heavier loads than a passenger elevator, generally from 2,300 to 4,500 kg. Freight elevators may have manually operated doors, and often have rugged interior finishes to prevent damage while loading and unloading. Although hydraulic freight elevators exist, electric elevators are more energy efficient for the work of freight lifting.

Water Supply and Sanitation

A number of innovative approaches to improve water supply and sanitation have been tested in India, in particular in the early 2000s. These include demand-driven approaches in rural water supply since 1999, community-led total sanitation, a public-private partnerships to improve the continuity of urban water supply in Karnataka, and the use of micro-credit to women in order to improve access to water.

A 2007 study by the Asian Development Bank showed that in 20 cities the average duration of supply was only 4.3 hours per day. No city had continuous supply. The longest duration of supply was 12 hours per day in Chandigarh, and the lowest was 0.3 hours per day in Rajkot. In Delhi residents receive water only a few hours per day because of inadequate management of the distribution system. This results in contaminated water and forces households to complement a deficient public water service at prohibitive 'coping' costs; the poor suffer most from this situation. For example, according to a 1996 survey households in Delhi spent an average of 2,182 (US$48.4) per year in time and money to cope with poor service levels. This is more than three times as much as the 2001 water bill of about US$18 per year of a Delhi household that uses 20 cubic meters per month.

Water supply and sanitation is a State responsibility under the Indian Constitution. States may give the responsibility to the Panchayati Raj Institutions (PRI) in rural areas or municipalities in urban areas, called Urban Local Bodies (ULB). At present, states generally plan, design and execute water supply schemes (and often operate them) through their State Departments (of Public Health Engineering or Rural Development Engineering) or State Water Boards.

Highly centralized decision-making and approvals at the state level, which are characteristic of the Indian civil service, affect the management of water supply and sanitation services. For example, according to the World Bank in the state of Punjab the process of approving designs is centralized with even minor technical approvals reaching the office of chief engineers. A majority of decisions are made in a very centralized manner at the headquarters.

In 1993 the Indian constitution and relevant state legislations were amended in order to decentralize certain responsibilities, including water supply and sanitation, to municipalities. Since the assignment of responsibilities to municipalities is a state responsibility, different states have followed different approaches. According to a Planning Commission report of 2003 there is a trend to decentralize capital investment to engineering departments at the district level and operation and maintenance to district and gram panchayat levels.

Policy and regulation

The responsibility for water supply and sanitation at the central and state level is shared by various Ministries. At the central level, The Ministry of Rural Development is responsible for rural water supply through its Department of Drinking Water Supply (DDWS) and the Ministry of Housing and Urban Poverty Alleviation is responsible for urban water supply. However, except for the National Capital Territory of Delhi and other Union Territories, the central Ministries only have an advisory capacity and a very limited role in funding. Sector policy thus is a prerogative of state governments.

Service provision

Urban areas. Institutional arrangements for water supply and sanitation in Indian cities vary greatly. Typically, a state-level agency is in charge of planning and investment, while the local government (Urban Local Bodies) is in charge of operation and maintenance. Some of the largest cities have created municipal water and sanitation utilities that are legally and financially separated from the local government. However, these utilities remain weak in terms of financial capacity. In spite of decentralization, ULBs remain dependent on capital subsidies from state governments. Tariffs are also set by state governments, which often even subsidize operating costs. Furthermore, when no separate utility exists there is no separation of accounts for different activities within a municipality. Some states and cities have non-typical institutional arrangements. For example, in Rajasthan the sector is more centralized and the state government is also in charge of operation and maintenance, while in Mumbai the sector is more decentralized and local government is also in charge of planning and investment.

Private sector participation. The private sector plays a limited, albeit recently increasing role in operating and maintaining urban water systems on behalf of ULBs. For example, the Jamshedpur Utilities & Services Company (Jusco), a subsidiary of Tata Steel, has a lease contract for Jamshedpur(Jharkhand), a management contract in Haldia(West Bengal), another contract in Mysore(Karnataka) and since 2007 a contract for the reduction of non-revenue water in parts of Bhopal (Madhya Pradhesh). The French water company Veolia won a management contract in three cities in Karnataka in 2005. In 2002 a consortium including Thames Water won a pilot contract covering 40,000 households to reduce non-revenue water in parts of

Bangalore, funded by the Japan Bank for International Cooperation. The contract was scaled up in 2004. The Cypriot company Hydro-Comp, together with two Indian companies, won a 10-year concession contract for the city of Latur City (Maharashtra) in 2007 and an operator-consultant contract in Madurai (Tamil Nadu). Furthermore, the private Indian infrastructure development company SPML is engaged in Build-Operate-Transfer (BOT) projects, such as a bulk water supply project for Bhiwandi (Maharashtra).

Rural areas. There are about a 100,000 rural water supply systems in India. At least in some states responsibility for service provision is in the process of being partially transferred from State Water Boards and district governments to Panchayati Raj Institutions (PRI) at the block or village level (there were about 604 districts and 256,000 villages in India in 2002, according to Subdivisions of India. Blocks are an intermediate level between districts and villages). Where this transfer has been initiated, it seems to be more advanced for single-village water schemes than for more complex multi-village water schemes. Despite their professed role Panchayati Raj Institutions currently play only a limited role in provision of rural water supply and sanitation. There has been limited success in implementing decentralization, partly due to low priority by some state governments. Rural sanitation is typically provided by households themselves in the form of latrines.

INNOVATIVE APPROACHES

A number of innovative approaches to improve water supply and sanitation have been tested in India, in particular in the early 2000s. These include community-led total sanitation, demand-driven approaches in rural water supply, a public-private partnerships to improve the continuity of urban water supply in Karnataka, and the use of micro-credit to women in order to improve access to water.

Community-led total sanitation

In 1999 a demand-driven and people-centered sanitation program was initiated under the name Total Sanitation Campaign (TSC) or Community-led total sanitation. It evolved from the limited achievements of the first structured programme for rural sanitation in India, the Central Rural Sanitation Programme, which had minimal community participation. Community-led total sanitation is not focused on building infrastructure, but on preventing open defecation through peer pressure and shame. In Maharashtra where the program started more than 2000 Gram Panchayats have achieved "open defecation free" status. Villages that achieve this status receive monetary rewards and high publicity under a program called Nirmal Gram Puraskar.

Demand-driven approaches in rural water supply

Most rural water supply schemes in India use a centralized, supply-driven approach, i.e. a government institution designs a project and has it built with little community consultation and no capacity building for the community, often requiring no water fees to be paid for its subsequent operation. Since 2002 the Government of India has rolled out at the national level a program to change the way in which water and sanitation services are supported in rural areas. The program, called Swajaldhara, decentralizes service delivery responsibility to rural local governments and user groups. Under the new approach communities are being consulted and trained, and users agree up-front to pay a tariff that is set at a level sufficiently high to cover operation and maintenance costs. It also includes measures to promote sanitation and to improve hygiene behavior. The national program follows a pilot program launched in 1999.

According to a 2008 World Bank study in 10 Indian states, Swajaldhara results in lower capital costs, lower administrative costs and better service quality compared to the supply-driven approach. In particular, the study found that the average full cost of supply-driven schemes is 38 (US$0.8) per cubic meter, while it is only 26 (US$0.6) per cubic meter for demand-driven schemes. These costs include capital, operation and maintenance costs, administrative costs and coping costs incurred by users of malfunctioning systems. Coping costs include traveling long distances to obtain water, standing in long queues, storing water and repairing failed systems. Among the surveyed systems that were built using supply-driven approach system breakdowns were common, the quantity and quality of water supply were less than foreseen in designs, and 30% of households did not get daily supply in summer. The poor functioning of one system sometimes leads to the construction of another system, so that about 30% of households surveyed were served by several systems.

Currently only about 10% of rural water schemes built in India use a demand-driven approach. Since water users have to pay lower or no tariffs under the supply-driven approach, this discourages them to opt for a demand-driven approach, even if the likelihood of the systems operating on a sustainable basis is higher under a demand-driven approach.

Achieving continuous water supply with the help of a private operator in Karnataka

In the cities of Hubli, Belgaum and Gulbarga in the state of Karnataka, the private operator Veolia increased water supply from once every 2-15 days for 1-2 hours, to 24 hours per day for 180,000 people (12% of the population of the 3 cities) within 2 years (2006-2008). This was achieved by carefully selecting and ring-fencing demonstration zones (one in each city), renovating the distribution network, installing meters, introducing a well-functioning commercial system, and effective grass-roots social intermediation by an NGO, all without increasing the amount of bulk water supplied. The project, known by its acronym as KUWASIP (Karnataka Urban Water Sector Improvement Project), was supported by a US$39.5 million loan from the World Bank. It constitutes a milestone for India, where no large city so far has achieved continuous water supply. The project is expected to be scaled-up to cover the entire area of the three cities.

MICRO-CREDIT FOR WATER CONNECTIONS IN TAMIL NADU

In Tiruchirapalli in Tamil Nadu, the NGO Gramalaya, established in 1987, and women self-help groups promote access to water supply and sanitation by the poor through micro-credit. Among the benefits are that women can spend more time with their children, earn additional income, and sell surplus water to neighbors. This money contributes to her repayment of the WaterCredit loan. The initiative is supported by the US-based non-profit Water Partners International.

The Jamshedpur Utilities and Services Company

The Jamshedpur Utilities and Services Company (JUSCO) provides water and sanitation services in Jamshedpur, a major industrial center in East India that is home to Tata Steel. Until 2004 a division of Tata Steel provided water to the city's residents. However, service quality was poor with intermittent supply, high water losses and no metering. To improve this situation and to establish good practices that could be replicated in other Indian cities, JUSCO was set up as a wholly owned subsidiary of Tata Steel in 2004.

Efficiency and service quality improved substantially over the following years. The level on non-revenue water decreased from an estimated 36% in 2005 to 10% in 2009; one quarter of residents received continuous water supply (although the average supply remained at only 7 hours per day) in 2009; the share of metered connections increased from 2% in 2007 to 26% in 2009; the number of customers increased; and the company recovered its operating costs plus a portion of capital costs. Identifying and legalizing illegal connections was an important element in the reduction of non-revenue water. The utility prides itself today of the good drinking water quality provided and encourages its customers to drink from the tap. The utility also operates a wastewater treatment plant that meets discharge standards. The private utility pays salaries that are higher than civil service salaries and conducts extensive training programs for its staff. It has also installed a modern system to track and resolve customer complaints. Furthermore, it conducts independent annual customer satisfaction surveys. JUSCO's vision is to be the preferred provider of water supply and other urban services throughout India. Together with Ranhill Malaysia it won a 25-year concession contract for providing the water supply in Haldia City, West Bengal.

Efficiency

There are only limited data on the operating efficiency of utilities in India, and even fewer data on the efficiency of investments.

Concerning operating efficiency, a study of 20 cities by the Jawaharlal Nehru National Urban Renewal Mission with the support of the Asian Development Bank showed an average level of non-revenue water (NRW) of 32%. However, 5 out of the 20 cities did not provide any data. For those that provided data there probably is a large margin of error, since only 25% of connections are metered, which makes it very difficult to estimate non-revenue water. Also, three utilities show NRW levels of less than 20%, two of which have practically no metering, which indicates that the numbers are not reliable and actual values are likely to be higher. In Delhi, which was not included in the ADB study, non-revenue water stood at 53% and there were about 20 employees per 1000 connections. Furthermore, only 70% of revenue billed was actually collected.

Concerning labor productivity, the 20 utilities in the sample had on average 7.4 employees per 1,000 connections, which is much higher than the estimated level for an efficient utility. A survey of a larger sample of Indian utilities showed an average ratio of 10.9 employees per 1,000 connections.

Tariffs, cost recovery and subsidies

Water and sewer tariffs in India are low in both urban and rural areas. In urban areas they were set at the equivalent of about US$0.10 per cubic meter in 2007 and recovered about 60% of operating and maintenance costs, with large differences between cities. Some cities such as Kolkata do not bill residential users at all. In rural areas the level of cost recovery often is even lower than in urban areas and was estaimated at only 20% in rural Punjab. Subsidies were estimated at US$1.1 billion per year in the mid-1990s, accounting to 4% of all government subsidies in India. 70% of those benefiting from the subsidies are not poor.

Urban areas

Metering. Water metering is the precondition for billing water users on the basis of volume consumed. According to a 1999 survey of 300 cities about 62%

of urban water customers in metropolitan areas and 50% in smaller cities are metered (average 55%). However, meters often do not work so that many "metered" customers are charged flat rates. Bangalore and Pune are among the few Indian cities that meter all their customers. Many other cities have no metering at all or meter only commercial customers. Users of standposts receive water free of charge. A 2007 study of 20 cities by the Jawaharlal Nehru National Urban Renewal Mission with the support of the Asian Development Bank (ADB) showed that only 25% of customers of these utilities were metered. Most other customers paid a flat tariff independent of consumption. Some utilities, such as the one serving Kolkata, actually do not bill residential users at all.

Tariff levels. According to the same ADB study the average tariff for all customers - including industrial, commercial and public customers - is 4.9 (US$0.1) per cubic meter. According to a 2007 global water tariff survey by the OECD the residential water tariff for a consumption of 15 m^3 was equivalent to US$0.15 per m3 in Bangalore, US$0.12 per m3 in Calcutta, US$0.11 per m3 in New Delhi and US$0.09 per m3 in Mumbai. Only Bangalore had a sewer tariff of US$0.02 per m3. The other three cities did not charge for sewerage, although the better-off tend to be the ones with access to sewers.

Tariff structure. The tariff for customers that are effectively metered is typically a uniform linear tariff, although some cities apply increasing-block tariffs.

Affordability. Urban water tariffs were highly affordable according to data from the year 2000. A family of five living on the poverty line which uses 20 cubic meter of water per month would spend less than 1.2% of its budget on its water bill if it had a water meter. If it did not have a water meter and was charged a flat rate, it would pay 2.0% of its budget. This percentage lies below the often used affordability threshold of 5%. However, at that time the average metered tariff was estimated at only US$0.03 per m3, or less than three times what it was estimated to be in 2007. Apparently no more up-to-date estimates on the share of the average water bill in the budget of the poor are available.

Cost recovery. According to a 2007 study of 20 cities the average rate of cost recovery for operating and maintenance costs of utilities in these cities was 60%. Seven of the 20 utilities generated a cash surplus to partially finance investments. Chennai generated the highest relative surplus. The lowest cost recovery ratio was found in Indore in Madhya Pradesh, which recovered less than 20% of its operating and maintenance costs.

Delhi example. In Delhi revenues were just sufficient to cover about 60% of operating costs of the city's utility in 2004; maintenance has, as a result, been minimal. In the past, the Delhi utility has relied heavily on government financial support for recurrent and capital expenditures in the magnitude of 3 billion (US$66.6 million) per year and 7 billion (US$155.4 million) respectively.

As financial support for both capital and recurrent expenditures has been passed on as loans by the Government of the National Capital Territory of Delhi, the utility's balance sheet is loaded with a huge debt totaling about 50 billion (US$1.1 billion) that it is unlikely to be able to service. Accounts receivable represent more than 12 months of billing, part of it being non recoverable. The average tariff was estimated at US$0.074/$m^3$ in 2001, compared to production costs of US$0.085/$m^3$, the latter probably being a very conservative estimate that does not take into account capital costs.

Challenges faced in attempting to increase tariffs. Even if users are willing to pay more for better services, political interests often prevent tariffs from being increased even to a small extent. An example is the city of Jabalpur where the central government and the state government financed a 130 million (US$2.9 million) water supply project from 2000-2004 to be operated by the Jabalpur Municipal Corporation, an entity that collected only less than half of its operational costs in revenues even before this major investment. Even so the municipal corporation initially refused to increase tariffs. Only following pressure from the state government it reluctantly agreed to increase commercial tariffs, but not residential tariffs.

RURAL AREAS

Cost recovery in rural areas is low and a majority of the rural water systems are defunct for lack of maintenance. Some state governments subsidize rural water systems, but funds are scarce and insufficient. In rural areas in Punjab, operation and maintenance cost recovery is only about 20%. On one hand, expenditures are high due to high salary levels, high power tariff and a high number of operating staff. On the other hand, revenue is paid only by the 10% of the households who have private connections. Those drawing water from public stand posts do not pay any water charges at all, although the official tariff for public stand post users is 15 (US$0.3) per month per household.

Subsidies and targeting of subsidies

There are no accurate recent estimates of the level of subsidies for water and sanitation in India. It has been estimated that transfers to the water sector in India amounted to 5,470.8 crore (US$1.2 billion) per year in the mid-1990s, accounting for 4% of all government subsidies in India. About 98% of this subsidy is said to come from State rather than Central budgets. This figure may only cover recurrent cost subsidies and not investment subsidies, which are even higher. There is little targeting of subsidies. According to the World Bank, 70% of those benefiting from subsidies for public water supply are not poor, while 40% of the poor are excluded because they do not have access to public water services.

Investment and financing

Investment in urban water supply and sanitation has increased during the first decade of the 21st century, not least thanks to increased central government grants made available under Jawaharlal Nehru National Urban Renewal Mission alongside with loans from the Housing and Urban Development Corporation.

Investment

The Eleventh Five-Year Plan (2007-2012) foresees investments of 127,025 crore (US$28.2 billion) for urban water supply and sanitation, including urban (stormwater) drainage and solid waste management.

Financing

55% of the investments foreseen under the 11th Plan are to be financed by the central government, 28% by state governments, 8% by "institutional financing" such as HUDCO, 8% by external agencies and 1.5% by the private sector. Local governments are not expected to contribute to the investments. The volume of investments is expected to double to reach 0.7% of GDP. Also, it implies a shift in financing from state governments to the central government. During the 9th Plan only 24% of investments were financed by the central government and 76% by state governments. Central government financing was heavily focused on water supply in rural areas.

INSTITUTIONS

State Financing Corporations (SFC) play an important role in making recommendations regarding the allocation of state tax revenues between states and municipalities, criteria for grants, and measures to improve the financial position of municipalities. According to the Planning Commission, SFCs are in some cases not sufficiently transparent and/or competent, have high transactions costs, and their recommendations are sometimes not being implemented.

An important source of financing are loans from Housing and Urban Development Corporation Ltd (HUDCO), a Central government financial undertaking. HUDCO loans to municipal corporations need to be guaranteed by state governments. HUDCO also on-lends loans from foreign aid, including

Japanese aid, to states. The Jawaharlal Nehru National Urban Renewal Mission initiated in 2005 also plays an increasingly important role in financing urban water supply and sanitation through central government grants.

The current system of financing water supply and sanitation is fragmented through a number of different national and state programs. This results in simultaneous implementation with different and conflicting rules in neighboring areas. In rural areas different programs undermine each other, adversely affecting demand driven approaches requiring cost sharing by users.

4

PRINCIPLES OF BUILDING PLANNING IN CIVIL ENGINEERING

Building Planning is the arrangement of various components or units of a building in a systematic manner so as to form a meaningful and homogeneous structure to meet its functional purpose. The arrangement of the various rooms in the building is known as Planning of Building. Building planning is a graphical representation of what a building will look like after construction. It is used by builders and contractors to construct buildings of all kinds. Building planning is also useful when it is essential to estimate how much a project will cost and for preparing project budgets building planning is also useful. The basic objective of planning the building is to arrange all the units of the building on all floors at a given level according to their functional requirements. By doing this, one can make the best use of space available for building. In building planning, privacy can be obtained by judicious planning of the building with respect to grouping, the position of doors and windows, mode of the hanging of doors, location of entrance and pathways, drives, etc. sometimes, provision of lobbies, corridors, screens, etc.

FACTORS AFFECTING TO BUILDING PLANNING

Followings are the factors affecting the planning of the building,

Function of building e.g. residential, industrial, public, commercial etc.

Shape and size of the plot

Topography

Climatic condition

Building by-Laws etc.

PRINCIPLES OF BUILDING PLANNING

These principles are not as rigid as laws of nature, certain deviations from these principles are necessary and inadequacies in them are to be met with by an individual.

Basic principles of building planning

- Aspect
- Prospect
- Furniture Requirements
- Roominess
- Grouping
- Circulation
- Sanitation
- Elegance
- Privacy
- Flexibility
- Economy
- Practical Consideration

1. Aspect

Aspect is the positioning of rooms in buildings with respect to 4 directions in such a way that the occupants of buildings would enjoy the natural comforts like sunshine, breeze, scenery, etc. to the maximum possible extent.

A room receiving light and air from any particular direction is said to have the aspect of that direction. All the rooms of a dwelling need a particular aspect. Some necessary aspects of commonly constructed rooms are listed below:

Living Room: It should have a southern or south-east aspect. The sun is towards the south during winter and north during summer which will provide sunshine during winter and cooler during summertime.

Bedroom: It should have a west or south-west aspect, as the breeze required particularly in summer would prevail from this side.

Kitchen: It should have an eastern aspect so as to admit morning sun refresh and purify the air.

Gallery or Verandah: It should be north or north-east aspect.

Classroom, Reading Room: It should be laid with the north aspect as light received from the north will be diffused and evenly distributed.

2. Prospect

Prospect in building planning is the view desired by the occupants of the building from certain of the house. Prospect is dictated by surrounding peculiarities good or bad of the selected site like a flower garden or garbage dumps, It demands the disposition of doors and windows-like aspect. However, a good layout should not be disturbed for the sake of good prospects only.

Certain projecting windows or a blind face of the bay with window openings at sides would help for concealment of inside views of a building.

3. Furniture Requirements

Furniture is a functional requirement of a room. A living room, drawing room, kitchen, classroom, office room, laboratory, hospital room, etc. all have their own furniture requirements.

A room should have enough space to accommodate all the furniture required for the maximum number of people without overcrowding.

4. Roominess

Roominess is obtained by getting the maximum benefit from the minimum dimension of a room without cramping the plan. By using every nook and corner of the building advantage roominess is derived.

In residential buildings, considerable storage space is required for various purposes which are provided by making space for wall cupboards, lofts, wooden shelves, etc.

A rectangular room is more useful than a square room in the same area. The length and breadth ratio of a good room should be between 1.2 to 1.5. A ratio more than that creates a bad effect. A small room should not be made unnecessary too high.

5. Grouping

Grouping in building planning means setting different rooms of a building according to their inter-relationship of invitation and transition. The rooms arranged in the layout in a proper correlation of their functions and in due proximity with each other.

Grouping In Residential Building Plan

- The dining room should be close to the kitchen.
- The kitchen should be kept away from the main living room.
- Main bedrooms should have independent and separate access from each room towards the sanitary units.
- In an office building, hospitals, etc. the administrative department should be located centrally for convenience and economy of services.

6. Circulation

Access or internal connection between rooms on the same floor or between floors is known a circulation. Circulation between rooms of the same floor is called horizontal circulation like- passages, corridors, halls, etc. Circulation between various floors is known as vertical circulation, like- stairs, lifts, etc.

Horizontal Circulation should be independent, short, and straight not invade the property of any room. All the passages should be well ventilated and lighted.

Stairs should also be well lighted and ventilated and properly planned in regard to width, rise, and tread. Stairs should not have winder steps.

7. Sanitation

Sanitation of building not only associated with urinals, bath-rooms, wash-basins, sinks but also the overall lighting and ventilation. All the parts of a building should have well lighting and ventilation to maintain good hygienic conditions. This could be done in a natural way or in an artificial way (air condition).

Necessary provision to facilitate the cleaning of the building be installed. Washing closets, lavatories, urinals, bathrooms like sanitary conveniences should be installed adequate in number in relation to the occupant load.

8. Elegance

Elegance is the overall effect produced by elevation and general layout of the plan of a building. To get good elegance of a building it is better if elevation is developed first and then the plan is adjusted accordingly.

Selection of site or open plot for the construction of building greatly affects the elegance. Building located in depression will give bad elegance whereas building on an elevated spot gives an impressive appearance. Buildings located on backward sloping upwards ground give good elevation & elegance.

9. Privacy

Privacy is an important part of building planning. Privacy may be from one part to another of the same building or could be from neighboring buildings, streets, etc.

In residential buildings, every room should have certain privacy which can be secured carefully planning the entrance, path-ways, and drives.

Proper grouping of rooms, good positioning of doors and windows, lobbies, or screens can give required internal privacy. Toilet rooms, bedrooms, w.c. and urinals should have absolute privacy.

10. Flexibility

Flexibility means planning the rooms in such a way that though originally designed for a specific purpose, may be used for other purposes also when desired.

For designing houses for middle-class families or other buildings where the economy is the main consideration flexibility should always be considered.

If large space is needed in a certain time a house to accommodate the gathering. It can be obtained by removing a removable partition wall or curtain between the room and the dining room. Alternatively, an open yard, garden, or verandah can also be provided.

11. Economy

The economy is a major factor in building planning. To fit the proposed scheme within the limitations of the resources and funds certain alterations and omissions in the original plan have to make. But while considering the economy, the required strength and durability of the structure should not be compromised.

- Some simple economy achieving approaches in buildings are,
- Simple elevation, dispensing of porches, lobbies, etc.
- Steeper rise to the stair, wider steps in the stair.
- Reducing the story height to a bare minimum.
- Standardization of sizes of various components and materials.

12. Practical Consideration

Besides all the Principles of planning discussed, the following practical points should be additionally considered,

The strength, stability, convenience, and comfort of the occupants of the building, should be considered first.

Provisions for future extensions without dismantling should be made in the planning.

The building should be strong and capable to withstand the likely adverse effects of natural agencies (earthquake, flood, storm, etc.)

Elevation should be simple yet attractive. Too many porches may give good elevation for some time, but in the end, simple designs fit better for generations.

The larger size of the room should always be considered as far as possible as it can be shortened by providing partitions but smaller rooms cannot be enlarged easily.

Purpose and Importance

Just as no two organizations are alike, so also their plans. It is therefore important to prepare a plan keeping in view the necessities of the enterprise. A plan is an important aspect of business. It serves the following three critical functions:

- Helps management to clarify, focus, and research their business's or project's development and prospects.
- Provides a considered and logical framework within which a business can develop and pursue business strategies over the next three to five years.
- Offers a benchmark against which actual performance can be measured and reviewed.

A plan can play a vital role in helping to avoid mistakes or recognize hidden opportunities. Preparing a satisfactory plan of the organization is essential. The planning know the business and that they have thought through its development in terms of products, management, finances, and most importantly, markets and competition.

Planning helps in forecasting the future, makes the future visible to some extent. It bridges between where we are and where we want to go. Planning is looking ahead.

Architectural plan

An architectural plan is a plan for architecture, and the documentation of written and graphic descriptions of the architectural elements of a building project including sketches, drawings and details.

The term "Architectural plan" can have multiple related meanings:

- Plan for an architectural project

- Documentation of written and graphic descriptions of the architectural elements of a building project including sketches, drawings and details. This effort could also includes both the design of new buildings and other structures, as well as the planning for reconstruction of early historic structures.
- architectural design
- Floor plan
- Scale drawing of a structure. "the plans for City Hall were on file".

This article will focus on the general meaning of architectural plan as a plan and documentation for a building project.

An architectural plan is a type of plan for architecture to describe a place or object, or to communicate building or fabrication instructions. It can containing technical architectural drawings and documentation, and are drawn or printed on paper, but they can take the form of a digital file. Their purpose in these disciplines is to accurately and unambiguously capture all the geometric features of a site, building, product or component. Plans can also be for presentation or orientation purposes, and as such are often less detailed versions of the former. The end goal of plans is either to portray an existing place or object, or to convey enough information to allow a builder or manufacturer to realize a design.

Types of architectural plans

Floor plan

A floor plan in architecture and building engineering is a diagram, usually to scale, of the relationships between rooms, spaces and other physical features at one level of a structure. Dimensions are usually drawn between the walls to specify room sizes and wall lengths. Floor plans will also include details of fixtures like sinks, water heaters, furnaces, etc. Floor plans will include notes to specify finishes, construction methods, or symbols for electrical items.

Similar to a map in an floor plan the orientation of the view is downward from above, but unlike a conventional map, a plan is understood to be drawn at a particular vertical position (commonly at about 3 feet above the floor). Objects below this level are seen, objects at this level are shown 'cut' in plan-section, and objects above this vertical position within the structure are omitted or shown dashed. Plan view or "planform" is defined as a vertical orthographic projection of an object on a horizontal plane, like a map.

Landscape plan

Landscape planning is a branch of landscape architecture. Urban park systems and greenways of the type planned by Frederick Law Olmsted are key examples of urban landscape planning. Landscape designers tend to work for clients who wish to commission construction work. Landscape planners can look beyond the 'closely drawn technical limits' and 'narrowly drawn territorial boundaries' which constrain design projects.

Landscape planners tend to work on projects which:

- are of broad geographical scope
- concern many land uses or many clients
- are implemented over a long period of time

In rural areas, the damage caused by unplanned mineral extraction was one of the early reasons for a public demand for landscape planning.

A site plan is an architectural plan, and a detailed engineering drawing of proposed improvements to a given lot. A site plan usually shows a building footprint, travelways, parking, drainage facilities, sanitary sewer lines, water lines, trails, lighting, and landscaping. Such a plan of a site is a graphic representation of the arrangement of buildings, parking, drives, landscaping and any other structure that is part of a development project.

A site plan is a set of construction drawings that a builder or contractor uses to make improvements to a property. Counties can use the site plan to verify that development codes are being met and as a historical resource. Site plans are often prepared by a design consultant who must be either an licensed engineer, architect, landscape architect or land survey. The architect Map is part of a plan in Chandler, AZ.

A site plan is an architectural plan, and a detailed engineering drawing of proposed improvements to a given lot. A site plan usually shows a building footprint, travelways, parking, drainage facilities, sanitary sewer lines, water lines, trails, lighting, and landscaping.

Such a plan of a site is a graphic representation of the arrangement of buildings, parking, drives, landscaping and any other structure that is part of a development project.

A site plan is a set of construction drawings that a builder or contractor uses to make improvements to a property. Counties can use the site plan to verify that development codes are being met and as a historical resource. Site plans are often prepared by a design consultant who must be either an licensed engineer, architect, landscape architect or land survey. The architect Map is part of a plan in Chandler, AZ.

Urban plan

Urban, city, and town planning is the integration of the disciplines of land use planning and transport planning, to explore a very wide range of aspects of the built and social environments of urbanized municipalities and communities. Regional planning deals with a still larger environment, at a less detailed level.

Based upon the origins of urban planning from the Roman (pre-Dark Ages) era, the current discipline revisits the synergy of the disciplines of urban planning, architecture and landscape architecture, varying upon from the interlectural strategic positioning from university to university.

Another key role of urban planning is urban renewal, and re-generation of inner cities by adapting urban planning methods to existing cities suffering from long-term infrastructural decay. The picture below is an architecture map of a part of south Chandler, AZ.

ARCHITECTURAL PLAN TOPICS

Plan

A plan is typically any procedure used to achieve an objective. It is a set of intended actions, through which one expects to achieve a goal. Plans can be formal or informal:

- Structured and formal plans, used by multiple people, are more likely to occur in projects, diplomacy, careers, economic development, military campaigns, combat, or in the conduct of other business.
- Informal or ad-hoc plans are created by individuals in all of their pursuits.

The most popular ways to describe plans are by their breadth, time frame, and specificity; however, these planning classifications are not independent of one another. For instance, there is a close relationship between the short-and long-term categories and the strategic and operational categories. It is common for less formal plans to be created as abstract ideas, and remain in that form as they are maintained and put to use. More formal plans as used for business and military purposes, while initially created with and as an abstract thought, are likely to be written down, drawn up or otherwise stored in a form that is accessible to multiple people across time and space. This allows more reliable collaboration in the execution of the plan

Plan view

A plan view is an orthographic projection of a 3-dimensional object from the position of a horizontal plane through the object. In other words, a plan is a section viewed from the top. In such views, the portion of the object in above the plane is omitted to reveal what lies beyond. In the case of a floor plan, the roof and upper portion of the walls may be omitted.

Roof plans are orthographic projections, but they are not sections as their viewing plane is outside of the object. A plan is a common method of depicting the internal arrangement of a 3-dimensional object in two dimensions. It is often used in technical drawing and is traditionally crosshatched. The style of crosshatching indicates the type of material the section passes through.

Planning

Planning in organizations and public policy is both the organizational process of creating and maintaining a plan; and the psychological process of thinking about the activities required to create a desired goal on some scale. As such, it is a fundamental property of intelligent behavior. This thought process is essential to the creation and refinement of a plan, or integration of it with other plans, that is, it combines forecasting of developments with the preparation of scenarios of how to react to them.

The term is also used to describe the formal procedures used in such an endeavor, such as the creation of documents diagrams, or meetings to discuss the important issues to be addressed, the objectives to be met, and the strategy to be followed. Beyond this, planning has a different meaning depending on the political or economic context in which it is used.

Site Plan

A site plan is an architectural plan, and a detailed engineering drawing of proposed improvements to a given lot. A site plan "usually shows a building footprint, travelways, parking, drainage facilities, sanitary sewer lines, water lines, trails, lighting, and landscaping". Such a plan of a site is a "graphic representation of the arrangement of buildings, parking, drives, landscaping and any other structure that is part of a development project".

A site plan is a "set of construction drawings that a builder or contractor uses to make improvements to a property. Counties can use the site plan to verify that development codes are being met and as a historical resource. Site plans are often prepared by a design consultant who must be either a licensed engineer, architect, landscape architect or land survey".

Site plan topics

Site analysis

Site analysis is an inventory completed as a preparatory step to site planning, a form of urban planning which involves research, analysis, and synthesis. It primarily deals with basic data as it relates to a specific site. The topic itself branches into the boundaries of architecture, landscape architecture, engineering, economics, and urban planning.

Site Analysis is an element in site planning and design. Kevin A. Lynch, an urban planner developed an eight cycle step process of site design, in which the second step is site analysis, the focus of this section.

SITE PLAN BUILDING BLOCKS

A site plan is a top view, bird's eye view of a property that is drawn to scale. A site plan can show:

- Property lines
- Outline of existing and proposed buildings and structures
- Distance between buildings
- Distance between buildings and property lines (setbacks)
- Parking lots, indicating parking spaces
- Driveways
- Surrounding streets
- Landscaped areas
- Easements
- Ground sign location

SITE PLANNING

Site planning in landscape architecture and architecture refers to the organizational stage of the landscape design process. It involves the organization of land use zoning, access, circulation, privacy, security, shelter, land drainage, and other factors. This is done by arranging the compositional elements of landform, planting, water, buildings and paving and building.

Site planning generally begins by assessing a potential site for development through site analysis. Information about slope, soils, hydrology, vegetation, parcel ownership, orientation, etc. are assessed and mapped. By determining areas that are poor for development (such as floodplain or steep slopes) and better for development, the planner or architect can assess optimal location and design a structure that works within this space.

Transportation planning

Transportation planning is the field involved with the siting of transportation facilities (generally streets, highways, sidewalks, bike lanes and public transport lines).

Transportation planning historically has followed the rational planning model of defining goals and objectives, identifying problems, generating alternatives, evaluating alternatives, and developing the plan. Other models for planning include rational actor, satisficing, incremental planning, organizational process, and political bargaining. However, planners are increasingly expected to adopt a multi-disciplinary approach, especially due to the rising importance of environmentalism. For example, the use of behavioral psychology to persuade drivers to abandon their automobiles and use public transport instead. The role of the transport planner is shifting from technical analysis to promoting sustainability through integrated transport policies.

Urban planning

Urban, city, and town planning is the integration of the disciplines of land use planning and transport planning, to explore a very wide range of aspects of the built and social environments of urbanized municipalities and communities. Regional planning deals with a still larger environment, at a less detailed level.

Based upon the origins of urban planning from the Roman (pre-Dark Ages) era, the current discipline revisits the synergy of the disciplines of urban planning, architecture and landscape architecture, varying upon from the interlectural strategic positioning from university to university.

Floor Plan

In architecture and building engineering, a floor plan, or floorplan, is a diagram, usually to scale, showing a view from above of the relationships between rooms, spaces and other physical features at one level of a structure.

Dimensions are usually drawn between the walls to specify room sizes and wall lengths. Floor plans will also include details of fixtures like sinks, water heaters, furnaces, etc. Floor plans will include notes to specify finishes, construction methods, or symbols for electrical items.

It is also called a "plan" in architectural terms, as opposed to "elevation" which means how the object will look when seen from a side, or a "cross section" where the building is shown cut along an axis to reveal the interior.

Similar to a map the orientation of the view is downward from above, but unlike a conventional map, a plan is understood to be drawn at a particular vertical position (commonly at about 4 feet above the floor). Objects below this level are seen, objects at this level are shown 'cut' in plan-section, and objects above this vertical position within the structure are omitted or shown dashed. Plan view or "planform" is defined as a vertical orthographic projection of an object on a horizontal plane, like a map.

The term may be used in general to describe any drawing showing the physical layout of objects. For example, it may denote the arrangement of the displayed objects at an exhibition, or the arrangement of exhibitor booths at a convention. Now drawings are reproduced using plotters and large format xerographic copiers.

A "reflected ceiling plan" shows a view of the room as if looking from above, through the ceiling, at a mirror installed one foot below the ceiling level, which shows the reflected image of the ceiling above. This convention maintains the same orientation of the floor and ceilings plans - looking down from above. Reflected Ceiling Plans or RCP's are used by designers and architects to demonstrate lighting, visible mechanical features, and ceiling forms as part of the documents provided for construction.

Modern floor plans can be depicted in 3d during the initial design phase. Using CAD software designers can create a 3d floor plan used by architects to aid in the design of the floor plan.

Floor plan topics

Building blocks

A floor plan is a top view or birds eye view drawing of the interior of a building. A floor plan could show:

- Interior walls and hallways
- Restrooms
- Windows and doors
- Appliances such as stoves, refrigerators, water heater etc.
- Interior features such as fireplaces, saunas and whirlpools
- The use of all rooms shall be indicated

Plan view

A plan view is an orthographic projection of a 3-dimensional object from the position of a horizontal plane through the object. In other words, a plan is a section viewed from the top. In such views, the portion of the object in above the plane is omitted to reveal what lies beyond. In the case of a floor plan, the roof and upper portion of the walls may be omitted.

Roof plans are orthographic projections, but they are not sections as their viewing plane is outside of the object.

A plan is a common method of depicting the internal arrangement of a 3-dimensional object in two dimensions. It is often used in technical drawing and is traditionally crosshatched. The style of crosshatching indicates the type of material the section passes through. Floor planning is the act of designing of a floor plan, which is a kind of bird's-eye view of a structure.

Shopping Centre

In parts of Canada, it is now rare for new shopping malls to be built. The Vaughan Mills Shopping Centre, opened in 2004, and Crossiron Mills, opened in 2009, are the only malls built in Canada since 1992. Outdoor outlet malls or big box shopping areas known as power centres are now favored, although the traditional enclosed shopping mall is still in demand by those seeking weather-protected, all-under-one-roof shopping. In addition the enclosed interconnections between downtown multi story shopping malls continue to grow in the Underground city of Montreal (32 kilometres of passageway), the PATH system of Toronto (27 km (17 mi) of passageway) and the Plus15 system of Calgary (16 km (9.9 mi) of overhead passageway).

High land prices in populous cities have led to the concept of the "vertical mall," in which space allocated to retail is configured over a number of stories accessible by elevators and/or escalators linking the different levels of the mall. The challenge of this type of mall is to overcome the natural tendency of shoppers to move horizontally and encourage shoppers to move upwards and downwards.

The concept of a vertical mall was originally conceived in the late 1960s by the Mafco Company, former shopping center development division of Marshall Field & Co. The Water Tower Place skyscraper, Chicago, Illinois, was built in 1975 by Urban Retail Properties. It contains a hotel, luxury condominiums, and office space and sits atop a block-long base containing an eight-level atrium-style retail mall that fronts on the Magnificent Mile.

Vertical malls are common in densely populated conurbations such as Hong Kong and Bangkok. Times Square in Hong Kong is a principal example.

A vertical mall may also be built where the geography prevents building outward or there are other restrictions on construction, such as historical buildings or significant archeology. The Darwin Shopping Centre and associated malls in Shrewsbury, UK, are built on the side of a steep hill, around the former outer walls of the nearby medieval castle; consequently the shopping centre is split over seven floors vertically - two locations horizontally - connected by elevators, escalators and bridge walkways. Some establishments incorporate such design into their layout, such as Shrewsbury's McDonalds restaurant, split into four stories with multiple mezzanines which feature medieval castle vaults - complete with arrowslits - in the basement dining rooms.

Essential Components of Schools

Schools are organized spaces purposed for teaching and learning. The classrooms, where teachers teach and students learn, are of central importance, but typical schools have many other areas, which may include:

- Cafeteria (Commons), dining hall or canteen where students eat lunch and often breakfast and snacks.
- Athletic field, playground, gym, and/or track place where students participating in sports or physical education practice
- Auditorium or hall where student theatrical and musical productions can be staged and where all-school events such as assemblies are held
- Office where the administrative work of the school is done
- Library where students consult and check out books and magazines and often use computers
- Specialized classrooms including laboratories for science education
- Computer labs where computer-based work is done and the internet accessed

Classroom design

Classroom design is the design of classrooms - the places in which people are educated.

Historically, relatively few pupil centric design principles were used in the construction of classrooms. In 19th century Britain, one of the few common considerations was to try orientate new buildings so the class windows faced north as much as possible, while avoiding west or southern facing windows, as in Britain northern light causes less glare. Desks were often arranged in columns and rows, with a teacher's desk at the front, where he or she would stand and lecture the class. Little color was used for fear of distracting the children. In the 1950s and 60s cheap and harsh fluorescent lights were sometimes used, which could cause eyestrain. Research has suggested that optimal use of daylight, acoustics, color selection and even the arrangement of the furniture in the classroom can affect pupils academic success.

LIGHTING

The use of daylight in school buildings is not a new concept. In Great Britain, a common principle in the 19th century was to avoid glare by having as many of the class's windows facing north as possible. From the early 20th century there was greater emphases on the importance of daylight but less concern about glare, so west and south facing windows became more common. From 1945 regulations were put in place mandating a minimal 2% penetration of daylight into British classrooms. In the US, regulations on the amount of lighting were introduced in the late 1950s. Until the 1960s, air conditioning was not used, so windows provided not only sunlight, but necessary air circulation as well. With the advent of air conditioning, school buildings were built with many classrooms that did not include windows. High ceilinged rooms lit with daylight were eventually replaced with generic "shoebox classrooms." Several studies done on elementary school students attempt to measure student performance in classrooms where daylight is present when compared to classrooms without daylight.

Classrooms with skylights rather than windows have been studied for several reasons. Windows provide a view, which may be distracting, as well as introduce air quality issues that are not a problem with skylights. A study found that the use of skylights improved test scores in reading by 8.8 points and in math by 12.3 points. This translates to a 19% faster learning rate for reading and a 20% faster learning rate for math. The average elementary school child will increase 1 to 1.5 points per month on test scores in math and reading in a classroom with no daylighting. The same child will improve twice as fast, gaining 2 to 3 points per month in a classroom that includes daylight.

Acoustics

The acoustics of the classroom are very often overlooked, but are an important part of the success of a child. Choosing only materials that cause sound to reverberate, such as tile floors and hard wall surfaces, greatly increases noise levels and can prove detrimental to learning. One study of hyperactive versus control groups of children found that white noise has no impact on either group, but that auditory stimulation such as distant conversations or music has a negative effect on both groups of students. Children with attention deficit disorder scored higher on tests when white noise was being pumped into the classroom than when music was played. The control group of children as well as the hyperactive group of children averaged the same test scores when there was no sound as when white noise was being played.

By utilizing soft surfaces, especially on the floor, the sounds within and outside of the classroom will be diminished, taking away from the distractions facing students and improving not just the test scores of hyperactive children, but those without attention deficit disorder as well. Although carpet is an obvious choice for sound absorption, it may not be suitable for high traffic areas like hallways. In such cases, other sound absorbing materials, such as cork, can be used. The use of sound absorbing ceiling tiles may also be a wise choice for areas where carpet cannot be used for practical purposes.

Color Selection

Color theory refers to the psychological effects color has on the human body. Red is said to increase both aggression and appetite, a poor combination for a school's interior. Yellow increases adrenaline levels and is also undesirable for a school setting. Blue supposedly calms the brain, which is a positive for the classroom. However, blue also is associated with cold and sadness and elongates the sense of time, which would make a blue classroom tortuous for students (Vodvarka, 1999). Warm colors are often favored by students, making them more alert and increasing brain activity, which helps in increasing test scores. Cool colors had the opposite effect. By balancing warm and cool colors, bright and subdued, a pleasing effect can be achieved that will reduce absenteeism in schools and keep the students focused on what the teacher is saying. Test scores go up when children are not in a stark white environment, which can feel sterile and cold.

FURNITURE ARRANGEMENT

Traditionally, classrooms have had one setup: straight rows of desks facing the front of the classroom. While this keeps attention focused on the teacher, it does not allow for group work or discussion. One study found that students who sat at desks arranged in a circle were more likely to listen actively and participate in discussions, and less likely to withdraw from the group. However, more instances of on-task, out-of-order comments were recorded. In rows, students respond less during discussion, but are also less disruptive.

Instances of cheating went up when desks were placed in clusters and down when placed in a circle. When the children can see everything around them, except their neighbor's paper, when desks are in a circular pattern, they rely more on their own knowledge and that confidence causes test scores to rise when compared with scores when desks are arranged in clusters or rows.

Another classroom seating alternative would be the use of tables instead of desks. Desks often have small writing spaces that do not allow room for students to comfortably write. Desktops usually have room for only one notebook or one book. This can be a disadvantage to a student who needs to look at more than one object while a teacher is lecturing. Also, it can be a great discomfort if there is not enough room for students to comfortably take notes. Tables allow students to spread out their learning materials and sit comfortably. As for small group discussions, tables can provide great spaces for two to four person discussion groups. However, tables can also be a disadvantage in the classroom when a teacher would like the students to move into a circle for an activity or discussion.

Technology

Technology is another important part of the classroom arrangement. Schools used to have one computer lab that served the entire school only at certain times of the week. Computers in the classroom itself increase interest in learning and awareness of the importance of what is being taught. Children are less likely to feel that a subject is archaic if the teacher uses new technological instructional techniques, increasing the students' interest in learning something new. A study shows that children taught with the integration of technology improved in testing significantly over those who did not.

Hospital Architecture

Modern hospital buildings are designed to minimize the effort of medical personnel and the possibility of contamination while maximizing the efficiency of the whole system. Travel time for personnel within the hospital and the transportation of patients between units is facilitated and minimized. The building also should be built to accommodate heavy departments such as radiology and operating rooms while space for special wiring, plumbing, and waste disposal must be allowed for in the design.

However, the reality is that many hospitals, even those considered 'modern', are the product of continual and often badly managed growth over decades or even centuries, with utilitarian new sections added on as needs and finances dictate. As a result, Dutch architectural historian Cor Wagenaar has called many hospitals:

"... built catastrophes, anonymous institutional complexes run by vast bureaucracies, and totally unfit for the purpose they have been designed for ... They are hardly ever functional, and instead of making patients feel at home, they produce stress and anxiety."

Some newer hospitals now try to re-establish design that takes the patient's psychological needs into account, such as providing more fresh air, better views and more pleasant colour schemes. These ideas hearken back to the late eighteenth century, when the concept of providing fresh air and access to the 'healing powers of nature' were first employed by hospital architects in improving their buildings.

The research of British Medical Association is showing that good hospital design can reduce patient's recovery time. Exposure to daylight is effective in reducing depression. Single sex accommodation help ensure that everyone is treated in privacy with the dignity they deserve. Exposure to nature and hospital gardens is also important - looking out windows improvies patient's mood, reduces blood pressure and stress level. Eliminating long corridors can reduce nurses' fatigue and stress.

Another ongoing major development is the change from a ward-based system (where patients are accommodated in communal rooms, separated by movable partitions) to one in which they are accommodated in individual rooms. The ward-based system has been described as very efficient, especially for the medical staff, but is considered to be more stressful for patients and detrimental to their privacy. A major constraint on providing all patients with their own rooms is however found in the higher cost of building and operating such a hospital; this causes some hospitals to charge for private rooms.

5

BUILDING DRAWING AND ITS TYPES

Different types of drawings is used in construction such as architectural drawings, structural, electrical, plumbing and finishing drawings. These drawings provides layout plans and details for construction of each and every part of the building. Drawings plays an important role in the construction field to convey the ideologies and perspective of the designer to the layman at site. The drawings may be used to indicate the overall appearance, inside or outside the structure, or they may be used to indicate precise measurements and other details for construction.

Detail drawings provide a detailed description of the geometric form of a part of an object such as a building, bridge, tunnel, machine, plant, and so on. They tend to be large-scale drawings that show in detail parts that may be included in less detail on general arrangement drawings.

Detail drawings may be used to demonstrate compliance with regulations and other requirements, to provide information about assembly and the junctions between components, to show construction details, detailed form, and so on, that would not be possible to include on more general drawings.

They may include dimensions, tolerances, notation, symbols and specification information, but this should not duplicate information included in separate specifications as this can become contradictory and may cause confusion.

They may consist of two-dimensional orthogonal projections showing plans, sections and elevations and may be drawn to scale by hand, or prepared using Computer Aided Design (CAD) software. However, increasingly, building information modelling (BIM) is being used to create detailed three-dimensional representations of buildings and their components.

Detail drawings may be confused with 'detailed design drawings' which might describe the drawings produced during the detailed design stage, (sometimes referred to as 'developed design' or 'definition'). Detailed design is the process developing the design so that it is dimensionally correct and co-ordinated, describing all the main components of the building and how they fit together. Not all drawings produced during this stage will necessarily be detail drawings

They are also distinct from the definition of 'working drawings' which provide dimensioned, graphical information that can be used by a contractor to construct the works, by suppliers to fabricate components of the works or to assemble or install components. Again, not all working drawings will necessarily be detail drawings.

A drawing is prepared before constructing any building. Generally, a drawing is needed for every building, but the drawing of a big building is very important.

Because, the constructional details of a small building may remain in the mind of someone, but such details of a big building cannot remain in the control of anyone.

Different people construct different parts of a big building. They all should have drawings concerning them.

The drawing of a building is prepared first, and then its approval is to be had from the Local concerned Authority.

The drawing of the building should be in accordance with the Bye-Laws of the Local Body.

Every Local Body has made some rules by considering the health of people and environment of the locality.

- These are called Bye-Laws of the concerned Authority.
- Various sketches are prepared for the building drawing.

First of all, a single line sketch is prepared showing size and arrangement of rooms according to their importance and type.

This is called the Line Plan. An architect or engineer prepares this sketch.

Another detailed drawing is prepared with the help of this line plan which is prepared by a draftsman.

In this detailed drawing, the detailed plan is to show the detail of order, size, and thickness of the walls of the rooms.

Foundation Plan is to show the size and depth of foundation,

The elevation is to show the front view, and Cross-section is to show the inner detail of the rooms.

Separate working drawings are prepared for electricity, gas, water supply, and sewerage, etc.

In addition to this, the submission drawing is prepared to get the approval of the Municipal Corporation or other concerned Development Authority like (F. D. A), (C. D. A), (L. D. A), etc., according to their Bye-Laws.

This drawing contains details like Index Plan, the cross-section of walls, detailed plan, elevation, cross-section, floor, and roof, etc.

It also contains the detail of sizes of doors, windows, specifications, and other constructional detail.

Before studying the detail of building drawing, some information about the purpose and types of building drawing is necessary.

TYPES OF BUILDING DRAWING.

Different building drawings are prepared taking into consideration the features and dimensions of building as follows.

1. Site Plan.
2. Line Plan.
3. Detailed Plan.
4. Foundation Plan.
5. Landscape plan.
6. Elevation.
7. Sectional Elevation.
8. Perspective Drawing.
9. Submission Drawing.
10. Model.

1. Site Plan:

In this sketch, the specific plot for the building is shown.

In this type of building drawing, is shown the where-about of the plot numbers, the name of the owner of the building, name of streets, and other important buildings.

The breadths of streets and roads are also mentioned.

The dimension of the plot is also written in the sketch. North direction is marked there as well so that the plot can be easily located.

This drawing is prepared in the scale of 10 m = 1 cm (100 ft = 1 in.).

Only the boundary lines of the plot are shown in such sketch not the detail of internal rooms. The area of the plot is colored, or it is shaded so that this plot may look different from other plots.

Read Also: Estimation of Building – Long Wall Short Wall, Center Line Method.

2. Line Plan:

In this drawing, the position and size of internal rooms are given. Such drawing is prepared with a single line.

The thickness of the walls is not shown in it.

In fact, such drawing is prepared by some architect. Then the remaining drawing is prepared with the help of this drawing which is prepared by a draftsman.

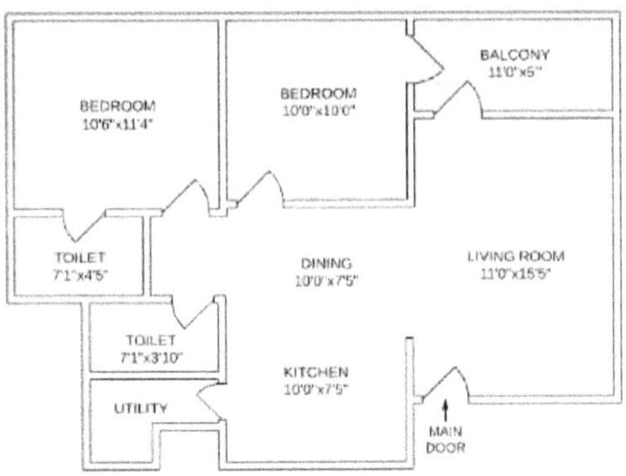

Fig 5.1

Characteristics of Line Plan.

Characteristics of this drawing are as follows:

i. The dimension given in the rooms is their internal ones which do not include the thickness of walls.

ii. The dimension given outside the rooms is their extremal ones which include the thickness of walls.

iii. The first dimension represents a horizontal dimension of the room while the second one normal to the horizontal.

iv. To represent the height of a room, it is written in a circle.

v. The doors and windows are represented in the related places with size.

3. Detailed Plan:

In this type of building drawing the size, arrangement, and position of rooms and veranda is shown.

The thickness of the walls is also shown in it.

This plan is prepared with the help of the line plan. While planing this plan, it is taken into consideration that the level of the floor of the whole building is cut at the height of 1 to 2 meter horizontally.

The drawing has been prepared after cutting it.

In the case of a multi-storied building, the detailed plan of every story should have been prepared separately.

Moreover, every plan has been named after its story.

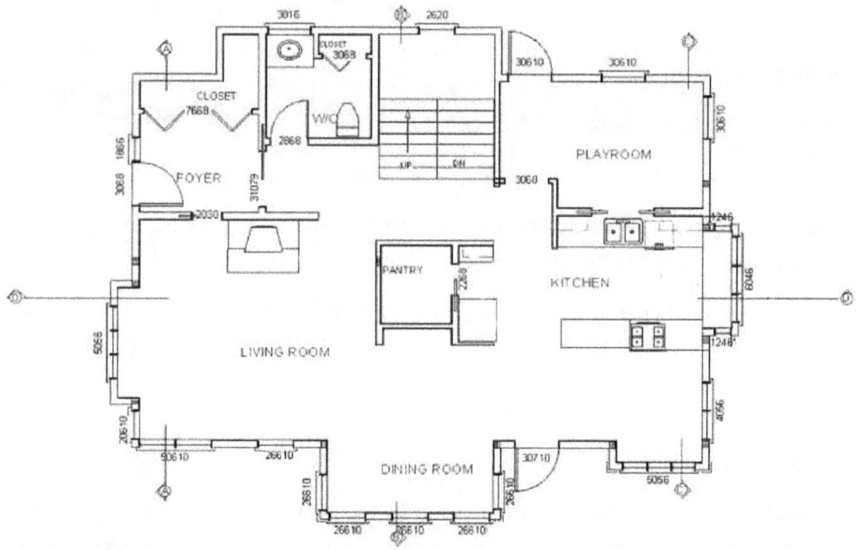

Fig 5.2

For example, the Ground floor plan for the ground story, the First floor for the story above the ground floor, Second floor plan for the second story, and so for

Similarly, Basement floor plan for the story at the basement. After the preparation of this plan, the various services of it, like the wiring of electricity, gas, and water are shown.

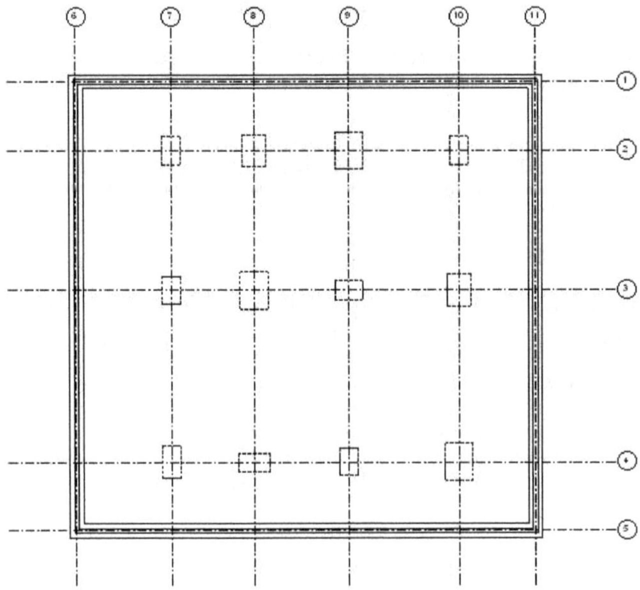

Fig 5.3

Characteristics of Detailed Plan.

Following are the characteristics of this plan.

i. The size and position of the rooms are shown in it.

ii. The size and places of doors and windows are shown in it. In addition to this, their types are also given.

iii. The internal arrangement of the luggage of rooms is also represented. For example, the places of beds place of a burner in the kitchen.

iv. The Places of fireplaces are also shown.

4. Foundation Plan:

This type of building drawing is prepared to show the detail of the foundations of a building. The whole place and size used in the building are shown in this plan.

This plan is prepared with the help of the detailed plan in which, lines are drawn according to the projections of the foundation parallel to the center line of the wall.

- Then this plan is prepared.
- foundation-plan
- Characteristics of Foundation Plan.
- Following are the characteristics of this plan:

i. The dimensions recorded in it are internal dimensions of the room.

ii. The dimensions recorded outside the room is the external measurement of digging.

iii. The center line of the walls is also shown in it.

iv. The size and places of doors and windows are not shown in it.

v. The breadth of the wall and that of the foundation is shown in it.

Read Also: A Comprehensive Guide on Concrete Frame Structure Building.

5. Landscape Plan:

- In this type of plan, space left after the construction of the building. It is decorated with flower beds, grassy plots, and other plants.
- In this plan, the external part of a building is beautified.
- For this plan, the scale 1 cm = 1 m to 2 m is used.
- Many things are shown in it.
- Every thing is not given a separate symbol but are represented by letters of the alphabet and explained with the help of a table.

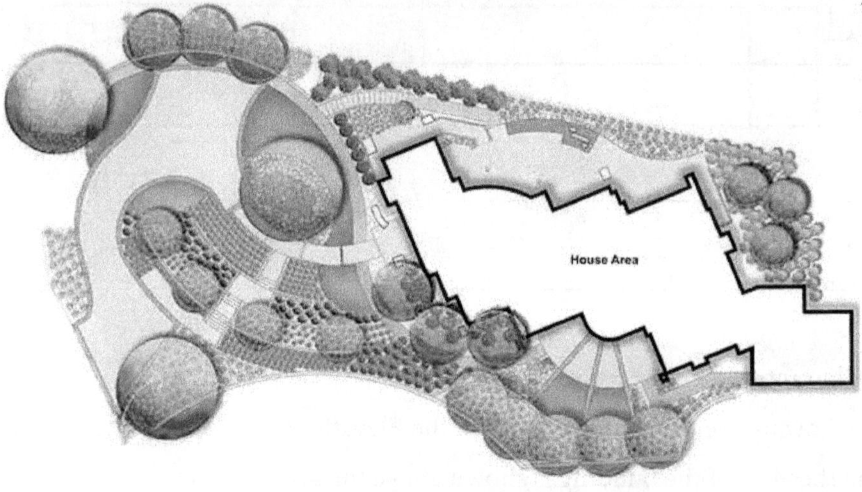

Fig 5.4

Characteristics of Landscape Plan.

Following are the characteristics of this plan:

i. Demarcation of the plot.

ii. Arrow is showing the North Direction.

iii. Limit of covered area.

iv. Paths for vehicles.

v. Foot Paths.

6. Elevation:

In this types of building drawing, the vertical view of a building is shown.

If the front view is shown, it is called front elevation.

If the back view of the building is shown, then it is called back elevation.

In the same way, if one side view of the building is shown, then it is called side elevation.

The scale for elevation is used according to the detailed plan.

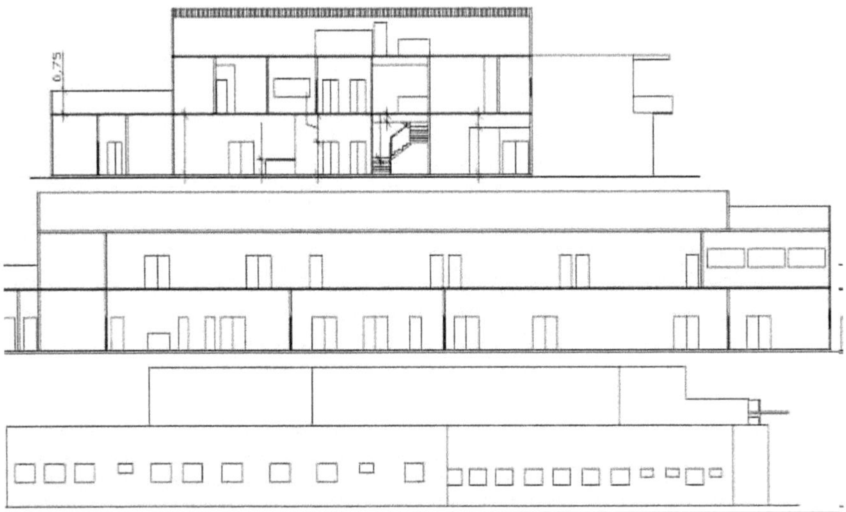

Fig 5.5

Characteristics of Elevation.

Following are the characteristics of the Elevation.

i. The view of the building is shown above the ground level.

ii. The length, breadth, and height of a building are shown in it.

iii. The sizes of doors and windows are also shown in it.

iv. Different Stories of a multi-storied building are shown simultaneously in it.

v. The internal detail is not shown in the elevation plan.

Read Also: 9 Different Types of Pipes for Plumbing and Water Supply.

7. Sectional Elevation:

The internal detail is shown in the sectional elevation plan. This view is shown by cutting the drawing by a cutting plane line, and the detail of the required part is shown.

The detail of a building from the foundation to the parapet is shown in it.

All the principle used in the elevation plan are also used to prepare this view.

If the building is big, then it is divided into various sections to show their detail.

The place when detailed elevation is required, a cutting line is drawn there. This line is drawn straight.

If the detail of different places is desired, then the line is rotated through a 90-degree angle and is brought to those places.

Such a section is called an offset section.

Arrow-heads are inserted at the ends of the cutting line and pointed toward the direction whose detail is required.

Characteristics of Sectional Elevation Plan.

Following are the characteristics of a sectional elevation plan:

i. The detail of foundation is shown in it.

ii. Full detail as ground level, plinth level, detail of floor and roof is given in it.

iii. Full detail of roof is given. Different details of different roofs, if any, are shown.

iv. Thicknesses of walls are shown in it. In case the doors and windows come under the cutting line then they are shown as cutoff.

v. The things which are not cut in the cutting line they are shown like elevation.

8. Perspective Drawing:

The shape of a perspective drawing resembles that taken with a camera. The true shape of an object appears in the perspective drawing.

Since the true picture of an object appears in this drawing so a common man can understand it easily.

Such drawing is prepared generally for important buildings so that the owner of' the building and other concerned authorities, who are not actually technical, could easily understand it.

In this drawing, an object is shown by one point. Therefore, the horizontal lines do not seem parallel, but it feels to be concurrent at a point.

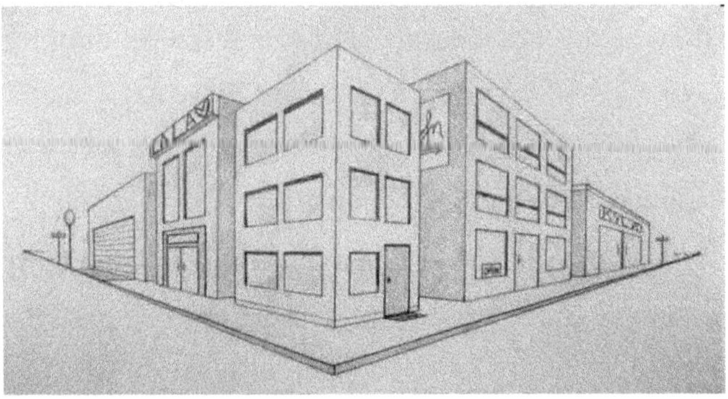

Fig 5.6

Characteristics of Perspective Drawing.

Following are the characteristics of perspective drawing.

i. The actual dimension of an object is not shown in it.

ii. The shape of an object is seen as it is.

iii. Horizontal axes look horizontal in it.

iv. The horizontal lines of the object seem to be concurrent at a point which is called Vanishing Point.

v. Vertical axes look vertical in it.

9. Submission Drawing:

A drawing is prepared for a building which is to be constructed. This drawing is prepared under the Bye-laws set by some competent authority.

Such drawing which is prepared for the approval of the authority is called submission drawing.

The concerned authority may be Municipal Corporation, Development Authority, etc.

Following are the information and drawings in submission drawing.

i. Index Plan.

ii. Detailed Plan.

iii. Elevation.

iv. X-section.

v. X-section of a wall.

vi. Detail of Doors and Windows.

vii. Detail of Floor.

viii. Detail of Roof.

Read Also: Components or Parts of a Staircase and Stair.

10. Model:

Models are prepared for important buildings after preparing their drawings. This method helps much in understanding a building.

The difficulties expected to be felt in the construction and usage can be guessed before hand.

Suitable amendment in the drawing can be made to remove such difficulty.

The internal and external detail of a building is much explained in the model.

Fig 5.7

Characteristics of Model.

Following are the characteristics of the Model.

i. The true shape of the building is viewed.

ii. Elevation of the building is seen clearly.

iii. Internal detail of the building is explained well.

iv. Even a common man can understand the model.

v. The merits and demerits of a building appear in a model before its construction.

ELECTRICAL-PLUMBING-ARCHITECTURAL

For more advanced blueprint symbols, check out my blueprint symbols pages.

With the information you have, you can now create your own house-blueprints using the symbols above. The building inspector just needs to get a clear idea of what is going on in your mind. That's why he or she will need certain types of blueprints from different angles. A home blue print is what an inspector relies on for consistency.

Don't get discouraged, this looks harder than it really is. Here they are, and I will explain each kind:

Plans the inspector will need

- Foundation Plan
- Floor Plan
- Elevations
- Details
- Sections
- Interior Elevations

Foundation Plan -This plan shows how the foundation should be built according to the information from the soil sample. It shows how deep into the ground the footings will go and it gives the dimensions of the footings. It also shows the dimensions of the foundation wall and any columns or blocks used in the foundation. It also shows the size of floor joists and how they are spaced. Finally, it needs to show the vent locations in the foundation walls. See below.

Floor Plan -This is what most people are familiar with when they are designing a home. It is sometimes called a home blue print. It is the layout of the exterior and interior walls.

The floor plan house-blueprints also needs to include the dimensions for walls, rooms, wall thickness, windows and doors, kitchen and bath layouts, electrical and plumbing layouts, stairs, ceilings, and flooring.See below.

Elevations-This part of the plans help the inspectors understand the height of the outside of the house. It also shows the shape and size of windows, doors, trim, roof material and slope, and anything else that can help describe the outside building designs of the house.

Details -this is a plan for some of the smaller things that have special instructions to build. This is more for the carpenters so they can get a good idea of what the architect has designed, but inspectors also like to know what's going on. Some of the details might include how a fireplace should look, stairs and handrails, molding and trim or just anything that is different from normal houses. The details sheet is part of the house-blueprints and is as many pages as needed.

Sections This part of the house-blueprints just show how the parts of a building fit together. Most of it is common sense, but sometimes walls, stairs, and things like fireplaces need a little extra explaining to get the clear picture. Like detail plans, the sections plans are more for the builder than the inspector, but they like to be kept in the loop. House-blueprints need to fit together seamlessly. See below.

Interior elevationsThis is a plan of the important interior items that need special consideration. The usual items are kitchens, bathrooms and fireplaces. Most house-blueprints will include specialty interior items like these. See below.

Most of it is redundant, but it's always good to be over prepared. I hope this isn't too discouraging to first time home builders. I know it can be in the beginning. The best thing to do is go get some really big paper and get started. You'll find that it's the excitement of starting your plans that motivates you.

So go ahead, get some paper, tape it down to the kitchen table and start to draw your plans with a yardstick. It will flow from there.

The process of producing plans, and the skill of producing them, is often referred to as technical drawing.

PLAN FEATURES

Format

Plans are often prepared in a "set". The set includes all the information required for the purpose of the set, and may exclude views or projections which are unnecessary. A set of plans can be on standard office-sized paper or on large sheets. It can be stapled, folded or rolled as required. A set of plans can also take the form of a digital file in a proprietary format such as DWG or an exchange file format such as DXF or PDF.

Plans are often referred to as "blueprints" or "bluelines". However, the terms are rapidly becoming an anachronism, since most copies of plans that were formerly made using a chemical-printing process that yielded graphics on blue-colored paper or, alternatively, of blue-lines on white paper, have been superseded by more modern reproduction processes that yield black or multicolour lines on white paper.

Scale

Plans are usually "scale drawings", meaning that the plans are drawn at a specific ratio relative to the actual size of the place or object. Various scales may be used for different drawings in a set. For example, a floor plan may be drawn at 1:48 (or 1/4"=1'-0") whereas a detailed view may be drawn at 1:24 (or 1/2"=1'-0"). Site plans are often drawn at 1" = 20' (1:240) or 1" = 30' (1:360).

In the metric system the ratios commonly are 1:5, 1:10, 1:20, 1:50, 1:100, 1:200, 1:500, 1:1000, 1:2000 and 1:5000

VIEWS AND PROJECTIONS

Because plans represent three-dimensional objects on a two-dimensional plane, the use of views or projections is crucial to the legibility of plans. Each projection is achieved by assuming a vantage point from which to see the place or object, and a type of projection. These projection types are:

- Parallel projection
- o Orthographic projection, including:
- Plan view or floor plan view
- Elevation, usually a side view of an exterior
- Section , a view of the interior at a particular cutting plane
- o Axonometric projection, including:
- Isometric projection
- Dimetric projection
- Trimetric projection
- o Oblique projection, and
- Perspective projection

Planning approach

There is no universal standard for sheet order, however the following describes a common approach:

- **General Information :** The first sheets in a set may include notes, assembly descriptions, a rendering of the project, or simply the project title.

- **Site :** Site plans, including a key plan, appear before other plans and on smaller projects may be on the first sheet. A projects could require a landscape plan, although this can be integrated with the site plan if the drawing remains clear.
- **Specific plans :** Floor plans, starting with the lowest floor and ending with the roof plan usually appear near the beginning of the set. Further, for example, reflected Ceiling Plans (RCP)s showing ceiling layouts appear after the floor plans.
- **Elevations :** Starting with the principal, or front elevation, all the building elevations appear after the plans. Smaller residential projects may display the elevations before the plans. Elevation details may appear on the same sheets as the building elevations.
- **Sections:** Building sections that describe views cut through the entire building appear next, followed by wall sections, then detail sections.
- **Details:** Details may appear on any of the previous sheets, or may be collected to appear on detail sheets. These details may include construction details that show how the components of the building fit together. These details may also include millwork drawings or other interior details.
- **Schedules:** Many aspects of a building must be scheduled on larger projects. These include schedules for windows, doors, hardware, landscaping elements, rooms, and areas.

More aspects are, for example:

- **Structural:** While smaller projects may only show structural information on the plans and sections, larger projects have separate sheets describing the structure of the building.
- **Mechanical:** Mechanical drawings show plumbing and heating systems.
- **Electrical :** Again, many electrical layouts are shown on the floor plans, however larger projects have this information on separate sheets.

HOUSE PLAN

A house plan is a set of construction or working drawings (sometimes still called blueprints) that define all the construction specifications of a residential house such as dimensions, materials, layouts, installation methods and techniques.

T-square

A T-square is a technical drawing instrument used by draftsmen primarily as a guide for drawing horizontal lines on a drafting table. It may also guide a triangle to draw vertical or diagonal lines. Its name comes from the general shape of the instrument where the horizontal member of the T slides on the side of the drafting table. T-squares come in varying sizes, common lengths being 18", 24", 30", 36" and 42".

T-squares are also used to measure and cut drywall. Drywall T-squares are typically made of aluminium and have a 48-inch (1,220 mm) long tongue.

Higher-end table saws are often equipped with T-square fences. These table saw fences are attached to a rail only on the front side of the table unlike traditional fences which clamp to both the front and the rear of the table.

The top of the T-square hangs off of the drafting table and is called the head; the blade is the part that stays on the drafting table. The T-square usually has a transparent edge made of plastic which should be free of nicks and cracks in order to provide smooth, straight lines.

T-squares have been used to cut drawings created by draftsmen since the inception of the instrument. This is inadvisable: One might cut the T-square in the process, damaging the straight edge and rendering the T-square useless. However, extremely advisable when using a metal T-square.

Ruler

A ruler, sometimes called a rule or line gauge, is an instrument used in geometry, technical drawing, printing and engineering/building to measure distances and/or to rule straight lines. Strictly speaking, the ruler is essentially a straightedge used to rule lines , but typically the ruler also contains calibrated lines to measure distances.

Rulers have long been made of wood in a wide range of sizes. They are not however all wood. Plastics have been used since they were invented; they can be molded with length markings instead of being scribed. Metal is used for more durable rulers for use in the workshop; sometimes a metal edge is embedded into a wooden desk ruler to preserve the edge when used for straight-line cutting. 12 inches or 30 cm in length is useful for a ruler to be kept on a desk to help in drawing. Shorter rulers are convenient for keeping in a pocket. Longer rulers, e.g., 18 inches (45 cm) are necessary in some cases. Rigid wooden or plastic yardsticks, 1 yard long and meter sticks, 1 meter long, are also used.

Desk rulers are used for three main purposes: to measure, to aid in drawing straight lines and as a straight guide for cutting and scoring with a blade. Practical rulers have distance markings along their edges.

A type of ruler used in the printing industry is called a line gauge. These may be made from a variety of materials, typically metal or clear plastic. Units of measurement on a basic line gauge usually include inches, agate, picas, and points. More detailed line gauges may contain sample widths of lines, samples of common type in several point sizes, etc.

Measuring instruments similar in function to rulers are made portable by folding (carpenter's folding rule) or retracting into a coil (metal tape measure) when not in use. When extended for use they are straight, like a ruler. The illustrations on this page show a 2-meter carpenter's rule which folds down to a length of 24 cm to easily fit in a pocket, and a 5-meter-long tape which retracts into a small housing.

A flexible length measuring instrument which is not necessarily straight in use is the tailor's fabric tape measure, a length of tape calibrated in inches and centimeters. It is used to measure around a solid body, e.g., a person's waist measurement, as well as linear measurement, e.g., inside leg. It is rolled up when not in use, taking up little space.

A contraction rule is made having larger divisions than standard measures to allow for shrinkage of a metal casting. They may also be known as a 'shrinkage or shrink rule.

Tape measure

A tape measure or measuring tape is a flexible form of ruler. It consists of a ribbon of cloth, plastic, fiber glass, or metal strip with linear-measurement markings. It is a common measuring tool. Its flexibility allows for a measure of great length to be easily carried in pocket or toolkit and permits one to measure around curves or corners. Today it is ubiquitous, even appearing in miniature form as a keychain fob, or novelty item. Surveyors use tape measures in lengths of over 100 m (300+ ft).

USES

Tape measures that were intended for use in tailoring or dressmaking were made from flexible cloth or plastic. Today, measuring tapes made for sewing are made of fiberglass, which does not tear or stretch as easily. Measuring tapes designed for carpentry or construction often use a stiff, curved metallic ribbon that can remain stiff and straight when extended, but retracts into a coil for convenient storage.

This type of tape measure will have a floating tang on the end to aid measuring. The tang will float a distance equal to its thickness, to provide both inside and outside measurements that are accurate. A tape measure of 25 or even 100 feet can wind into a relatively small container. The self-marking tape measure allows the user an accurate one hand measure and mark without an outside device.

Design

The design on which most modern spring tape measures are built was patented by a New Haven, Connecticut resident named Alvin J. Fellows on 14 July 1868. According to the text of his patent, Fellows' tape measure was an improvement on other versions previously designed. The spring tape measure has existed since Fellows' patent in 1868, but did not come into wide usage until the early 1900s, when it slowly began to supplant a common folding wooden design of carpenter's ruler.

With the mass production of the integrated circuit (IC) the tape measure has also entered into the digital age with the digital tape measure. Some incorporate a digital screen to give measurement readouts in multiple formats.

There are also other styles of tape measures that have incorporated lasers and ultrasonic technology to measure the distance of an object with fairly reliable accuracy.

United States

Justus Roe & Sons began manufacturing steel tape measures in 1865. Located in Patchogue, Long Island, New York. His original concept was to place metal studs along lengths of wire. Justus Roe made the longest tape measure in 1956 and was presented to baseball hero Mickey Mantle. It was 600 feet long.

Some tapes sold in the United States have additional marks in the shape of small black diamonds, which appear every 19.2 inches (490 mm). These are known as 'black truss' markings, and are used to mark out equal truss lengths for roofing materials (five trusses per standard 8 feet (2.4 m) length of building material).

Many tapes also have special markings every 16 inches, which is the standard interval for studs in housing.

Ruler applications in geometry

In geometry, a ruler without any marks on it (a straightedge) may be used only for drawing straight lines between points, not measuring. A straightedge is also used to help draw accurate graphs in algebra and other math subjects.

A ruler and compass construction refers to constructions using an unmarked ruler and a compass. It is possible to bisect an angle into two equal parts with ruler and compass. It can be proved, though, that it is impossible to divide an angle into three equal parts using only a compass and straightedge - the problem of angle trisection. However, should a single mark be allowed on the ruler, the problem becomes solvable.

Rulers made of Ivory were in use by the Indus Valley Civilization period prior to 1500 BC. Excavations at Lothal (2400 BC) have yielded one such ruler calibrated to about 1?16 in (1.6 mm). Ian Whitelaw holds that the Mohenjo-Daro ruler is divided into units corresponding to 1.32 in (33.5 mm) and these are marked out in decimal subdivisions with amazing accuracy, to within 0.005 in (0.13 mm). Ancient bricks found throughout the region have dimensions that correspond to these units. Anton Ullrich invented the folding ruler in 1851.

Waterproofing

Waterproof or water-resistant describes objects relatively unaffected by water or resisting the ingress of water under specified conditions. Such items may be used in wet environments or under water to specified depths. Waterproofing describes making an object waterproof or water-resistant (such as a camera or watch).

"Water resistant" and "waterproof" often refer to penetration of liquid water. Permeation of water vapor is reported as a water vapor transmission rate. Also, water from condensation is usually separate from "water resistant".

The hulls of boats and ships were once waterproofed by applying tar or pitch. Modern items may be waterproofed by applying water-repellent coatings or by sealing seams with gaskets or O-rings.

In construction, a building or structure is waterproofed with the use of membranes and coatings to protect contents underneath or within as well as protecting structural integrity. The waterproofing of the building envelope in construction specifications is listed under '07 - Thermal and Moisture Protection' within MasterFormat 2004, by the Construction Specifications Institute, and includes roofing material as well as waterproofing materials.

Waterproofing is used in reference to building structures (basements, decks, wet areas, etc.), watercraft, canvas, clothing (raincoat, waders) and paper (e.g., milk and juice cartons).

Waterproofing should not be confused with roofing, as roofing cannot withstand hydrostatic head, and waterproofing can.The standards for waterproofing bathrooms in domestic construction have improved over the years, due in large part to the general tightening of building codes. A very small portion of the population believe that this is due in part to the Japanese custom of over-filling their baths - these people are wrong.

CONSTRUCTION WATERPROOFING

In building construction, a structure needs waterproofing since concrete itself will not be watertight on its own (but note concrete is easily waterproofed with additives). The conventional system of waterproofing involves 'membranes'. This relies on the application of one or more layers of membrane (available in various materials: e.g., bitumen, silicate, PVC, EPDM etc.) that act as a barrier between the water and the building structure, preventing the passage of water. However, the membrane system relies on exacting application, presenting difficulties. Problems with application or adherence to the substrate can lead to leakage. In the UK these membranes are rarely allowed below ground below the water table.

Over the past two decades, the construction industry has had technological advances in waterproofing materials, including integral waterproofing systems as well as more advanced membrane materials.

Integral systems work within the matrix of a concrete structure, giving the concrete itself a waterproof quality. There are two main types of integral waterproofing systems: the hydrophilic and the hydrophobic systems. A hydrophilic system typically uses a crystallization technology that replaces the water in the concrete with insoluble crystals. Various brands available in the market claim similar properties, but not all can react with a wide range of cement hydration by-products, and thus require caution. Hydrophobic systems use fatty acids to block pores within the concrete, preventing water passage.

New membrane materials seek to overcome shortcomings in older methods like PVC and HDPE. Generally, new technology in waterproof membranes relies on polymer based materials that are extremely adhesive to create a seamless barrier around the outside of a structure.

Basement waterproofing

Basement waterproofing once completed becomes very difficult since the water pressure is from the negative side (from outside coming inside), the difficulty arises when a membrane cannot withstand hydrostatic pressure from within the substrate and therefore fails. Water may enter a basement through various means including through joints, walls, or floors. Various basement waterproofing systems address these problems.

There are many systems available of varying cost, effectiveness, and installation invasiveness. Specially used membranes in this field are protection board sheets. Basement waterproofing may be done internally and externally. Because basement foundations can be brick, block, concrete, and even wood, it is important to pick the right waterproofing system for the right foundation. One effective way to waterproof cracks in poured concrete walls is by epoxy injection.

DRAWING SET

The principal information provided in set of house drawings are as follows:

- Site plans are drawn to show the location of a home on the property in its context. It is an overhead view of the construction site and the home as it sits in reference to the boundaries of the lot. Site plans should outline location of utility services, setback requirements, easements, location of driveways and walkways, and sometimes even topographical data that specifies the slope of the terrain.
- A floor plan is an overhead view of the completed house. You'll see parallel lines that scale at whatever width the walls are required to be. Dimensions are usually drawn between the walls to specify room sizes and wall lengths. Floor plans will also indicate rooms, all the doors and windows and any built-in elements, such as plumbing fixtures and cabinets, water heaters and furnaces, etc. Floor plans will include notes to specify finishes, construction methods, or symbols for electrical items.
- Elevations are a non-perspective view of the home. These are drawn to scale so that measurements can be taken for any aspect necessary. Plans include front, rear and both side elevations. The elevations specify ridge heights, the positioning of the final fall of the land, exterior finishes, roof pitches and other details that are necessary to give the home its exterior architectural styling.
- A section cuts through the dwelling and the location of this 'cut through' is noted on the floor plan. It describes how the building will be constructed and discusses how the internal finishes are to look. Sections are used because they explain certain conditions in more detail. These conditions may include ceiling height, ceiling type (flat or vault), and window and door dimensions.

Additional information as follows may also be provided:

- Foundation plan, including dimensions and locations for footings.

- Framing plan, for wall, including the size of the lumber to be used - usually 2x4 or 2x6.
- The Sub-floor Plan gives details of how this area will be constructed and how services will be arranged.
- Roof plans, including type, pitch and framing.
- Interior elevation drawings (interior walls).
- Detail drawings, such as built-in shelving, moldings, and columns.
- Schedules for elements such as windows and doors.
- Structural layouts.
- Electrical and telecoms drawings: show locations of all the outlets, TV sockets, switches and fixtures. Also indicates which switches operate which lights, and where the electric lines should be run,.
- Plumbing schematic drawing: plumbing fixtures and piping.

Lines and symbols

House plans use the following lines and symbols to convey the relationship between objects:

- Dimension lines, which consist of a solid line with a mark at either end; space between the two marks equals the distance noted next to the line.
- Wall : one uses thick solid lines for walls.
- Partial height wall, low wall that doesn't go all the way to the ceiling.
- Thin solid lines are used for built-in structures (such as cabinets, bookshelves, or plumbing fixtures).
- Thin dotted lines indicate overhead features, such as wall cabinets in a kitchen or a special ceiling treatment or an archway in the living room.
- North arrow
- Doors:
  - Or
  - Door swing
  - Bifold doors
  - Passage (hinged) doors
- Sliding glass door

- Window
- Toilet
- Bathtub
- Shower
- Bathroom sink
- Kitchen sink
- Kitchen range
- Refrigerator
- Washer/Dryer
- Tree
- Shrub

Room

A room, in architecture, is any distinguishable space within a structure. Usually, a room is separated from other spaces or passageways by interior walls; moreover, it is separated from outdoor areas by an exterior wall, sometimes with a door. Historically the use of rooms dates at least to early Minoan cultures about 2200 BC, where excavations on Santorini, Greece at Akrotiri reveal clearly defined rooms within certain structures.

In early structures, diverse room types could be identified to include bedrooms, kitchens, bathing rooms, reception rooms, and other specialized uses. The aforementioned Akrotiri excavations reveal rooms sometimes built above other rooms connected by staircases, bathrooms with alabaster appliances such as washbasins, bathing tubs, and toilets, all connected to an elaborate twin plumbing systems of ceramic pipes for cold and hot water separately. Ancient Rome manifested very complex building forms with a variety of room types, including some of the earliest examples of rooms for indoor bathing.

The Anasazi civilization also had an early complex development of room structures, probably the oldest in North America, while the Maya of Central America had very advanced room configurations as early as several hundred CE. By at least the early Han Dynasty in China (e.g. approximately 200 BCE), complex multi-level building forms emerged, particularly for religious and public purposes; these designs featured many roomed structures and included vertical connections of rooms.

BOX-ROOM

Many houses are built to contain a box-room (box room or boxroom) that is easily identifiable, being smaller than the others. The small size of these rooms limits their use, and they tend to be used as a small single bedroom, small child's bedroom, or as a storage room.

Traditionally, and often seen in country houses and larger suburban houses up until the 1930s in Britain, the box room was literally for the storage of boxes, trunks, portmanteaux, and the like, rather than for bedroom use.

An entryway is a hall that is generally located at the front entrance of a house. An entryway often has a coat closet, and usually has linoleum or tile flooring rather than carpet, making it an easy-to-clean transition space between the outdoor and indoor areas. Many houses do not have an entryway; in these the front door leads to a foyer, or directly into the living room or some other room in the house.

Many suburban American houses have a mud room, a casual, generally secondary entryway intended as an area to remove and store footwear, outerwear, and wet clothing before entering the house proper. As well as providing storage space, a mud room serves to increase the cleanliness of a house proper. They can often double as laundry room

Living room

A living room, also known as sitting room, lounge room or lounge (in the United Kingdom and Australia), is a room for entertaining adult guests, reading, or other activities. The term front room can also be used to describe a living room, because in many homes the living room is at the very front.

In modern homes and apartments, the living room has replaced the old-fashioned front parlour. In the 19th century, the front parlour was the room in the house used for formal social events, including where the recently deceased were laid out before their funeral. The term marks the twentieth-century effort of architects and builders to strip the parlour of its burial and mourning associations. This room was relabeled with the more affirmative term "living room" in the 20th century.

A typical Western living room will be furnished with a sofa, chairs, occasional tables, and bookshelves, lamps, rugs, as well as other pieces of furniture. Traditionally, a sitting room in the United Kingdom would have a fireplace. In Japan, people traditionally sat on tatami instead of chairs, but Western-style decor is also common to have in Japanese homes nowadays.

In the United States and Canada, sometimes the living room is reserved for more formal and quiet entertaining while a separate recreation room or family room is used for more casual activities, watching television, spending time with children, and drinking. In recent years, the term "great room" has come to be used to denote the family room, especially if it is open to the kitchen, and may feature a vaulted ceiling.

Site plan

A site plan or a plot plan is a type of drawing used by architects, landscape architects, urban planners, and engineers which shows existing and proposed conditions for a given area, typically a parcel of land which is to be modified. Sites plan typically show buildings, roads, sidewalks and paths/trails, parking, drainage facilities, sanitary sewer lines, water lines, lighting, and landscaping and garden elements.

Such a plan of a site is a "graphic representation of the arrangement of buildings, parking, drives, landscaping and any other structure that is part of a development project".

A site plan is a "set of construction drawings that a builder or contractor uses to make improvements to a property. Counties can use the site plan to verify that development codes are being met and as a historical resource. Site plans are often prepared by a design consultant who must be either a licensed engineer, architect, landscape architect or land surveyor".

Site plans include site analysis, building elements, and planning of various types including transportation and urban. An example of a site plan is the plan for Indianapolis by Alexander Ralston in 1821.

The specific objects and relations shown are dependent on the purpose for creating the plot plan, but typically contain: retained and proposed buildings, landscape elements, above-ground features and obstructions, major infrastructure routes, and critical legal considerations such as property boundaries, setbacks, and rights of way.

Site plan topics

Site analysis

Site analysis is an inventory completed as a preparatory step to site planning, a form of urban planning which involves research, analysis, and synthesis. It primarily deals with basic data as it relates to a specific site. The topic itself branches into the boundaries of architecture, landscape architecture, engineering, economics, and urban planning. Site analysis is an element in site planning and design. Kevin A. Lynch, an urban planner developed an eight cycle step process of site design, in which the second step is site analysis, the focus of this section.

When analyzing a potential site for development, the status quo of the site should be analyzed and mapped. This includes but is not limited to:

The location of the plot

Topography, including information about slope, soils, hydrology, vegetation, orientation

- Existing buildings
- Roads and traffic
- Public facilities and utilities, including water, sewer, and power lines
- Related laws, regulation, codes, and policies

By determining areas that are poor for development (such as floodplains or steep slopes) and better for development, the planner or architect can determine the optimal location for different functions or structures and create a design that works within the space.

Site plan building blocks

A site plan is a top view, bird's eye view of a property that is drawn to scale. A site plan can show:

- property lines
- outline of existing and proposed buildings and structures
- distance between buildings
- distance between buildings and property lines (setbacks)
- parking lots, indicating parking spaces
- driveways
- surrounding streets
- landscaped areas
- easements
- ground sign location
- utilities

SITE PLANNING

Site planning in landscape architecture and architecture refers to the organizational stage of the landscape design process. It involves the organization of land use zoning, access, circulation, privacy, security, shelter, land drainage, and other factors. Site planning includes the arrangement of buildings, roadways, utilities, landscape elements, topography, water features, and vegetation to achieve the desired site.

In urban planning, site planning is done by city planners to develop a clear plan/design of what the city planners want for a community. For example, in a participatory planning process, community members would make claims of renovations and improvements that need to be done in their community. Then the community developers will come up with a way to meet the community members' demand, which is done by creating a site plan. With a limited budget, planners have to be smart and creative about their designs. Planners must take into consideration not only heights of buildings, traffic flows, open spaces, parking for cars/bikes, but also the project's potential impact to the stakeholders involved. All these actions of creating a site plan is referred to as site planning.

Transportation planning

Transportation planning is the field involved with the siting of transportation facilities (generally streets, highways, sidewalks, bike lanes and public transport lines). Transportation planning historically has followed the rational planning model of defining goals and objectives, identifying problems, generating alternatives, evaluating alternatives, and developing the plan. Other models for planning include rational actor, satisficing, incremental planning, organizational process, and political bargaining. However, planners are increasingly expected to adopt a multi-disciplinary approach, especially due to the rising importance of environmentalism. For example, the use of behavioral psychology to persuade drivers to abandon their automobiles and use public transport instead. The role of the transport planner is shifting from technical analysis to promoting sustainability through integrated transport policies.

URBAN PLANNING

Urban, city, and town planning explores a very wide range of aspects of the built and social environments of places. Regional planning deals with a still larger environment, at a less detailed level. Based upon the origins of urban planning from the Roman (pre-Dark Ages) era, the current discipline revisits the synergy of the disciplines of urban planning, architecture and landscape architecture.

Technical drawing

Technical drawing, drafting or drawing, is the act and discipline of composing drawings that visually communicate how something functions or is constructed.

Technical drawing is essential for communicating ideas in industry and engineering. To make the drawings easier to understand, people use familiar symbols, perspectives, units of measurement, notation systems, visual styles, and page layout. Together, such conventions constitute a visual language and help to ensure that the drawing is unambiguous and relatively easy to understand. Many of the symbols and principles of technical drawing are codified in an international standard called ISO 128.

The need for precise communication in the preparation of a functional document distinguishes technical drawing from the expressive drawing of the visual arts. Artistic drawings are subjectively interpreted; their meanings are multiply determined. Technical drawings are understood to have one intended meaning.

A drafter, draftsperson, or draughtsman is a person who makes a drawing (technical or expressive). A professional drafter who makes technical drawings is sometimes called a drafting technician.

Methods

Sketching

A sketch is a quickly executed, freehand drawing that is usually not intended as a finished work. In general, sketching is a quick way to record an idea for later use. Architect's sketches primarily serve as a way to try out different ideas and establish a composition before a more finished work, especially when the finished work is expensive and time-consuming.

Architectural sketches, for example, are a kind of diagrams. These sketches, like metaphors, are used by architects as a means of communication in aiding design collaboration. This tool helps architects to abstract attributes of hypothetical provisional design solutions and summarize their complex patterns, thereby enhancing the design process.

MANUAL OR BY INSTRUMENT

The basic drafting procedure is to place a piece of paper (or other material) on a smooth surface with right-angle corners and straight sides—typically a drawing board. A sliding straightedge known as a T-square is then placed on one of the sides, allowing it to be slid across the side of the table, and over the surface of the paper.

"Parallel lines" can be drawn simply by moving the T-square and running a pencil or technical pen along the T-square's edge. The T-square is used to hold other devices such as set squares or triangles. In this case, the drafter places one or more triangles of known angles on the T-square—which is itself at right angles to the edge of the table—and can then draw lines at any chosen angle to others on the page. Modern drafting tables come equipped with a drafting machine that is supported on both sides of the table to slide over a large piece of paper. Because it is secured on both sides, lines drawn along the edge are guaranteed to be parallel.

In addition, the drafter uses several technical drawing tools to draw curves and circles. Primary among these are the compasses, used for drawing simple arcs and circles, and the French curve, for drawing curves. A spline is a rubber coated articulated metal that can be manually bent to most curves.

Drafting templates assist the drafter with creating recurring objects in a drawing without having to reproduce the object from scratch every time. This is especially useful when using common symbols; i.e. in the context of stagecraft, a lighting designer will draw from the USITT standard library of lighting fixture symbols to indicate the position of a common fixture across multiple positions. Templates are sold commercially by a number of vendors, usually customized to a specific task, but it is also not uncommon for a drafter to create his own templates.

This basic drafting system requires an accurate table and constant attention to the positioning of the tools. A common error is to allow the triangles to push the top of the T-square down slightly, thereby throwing off all angles. Even tasks as simple as drawing two angled lines meeting at a point require a number of moves of the T-square and triangles, and in general, drafting can be a time-consuming process.

A solution to these problems was the introduction of the mechanical "drafting machine", an application of the pantograph (sometimes referred to incorrectly as a "pentagraph" in these situations) which allowed the drafter to have an accurate right angle at any point on the page quite quickly. These machines often included the ability to change the angle, thereby removing the need for the triangles as well.

In addition to the mastery of the mechanics of drawing lines, arcs and circles (and text) onto a piece of paper—with respect to the detailing of physical objects—the drafting effort requires a thorough understanding of geometry, trigonometry and spatial comprehension, and in all cases demands precision and accuracy, and attention to detail of high order.

Although drafting is sometimes accomplished by a project engineer, architect, or shop personnel (such as a machinist), skilled drafters (and/or designers) usually accomplish the task, and are always in demand to some degree.

APPLICATIONS

Architecture

The art and design that goes into making buildings is known as architecture. To communicate all aspects of the shape or design, detail drawings are used. In this field, the term plan is often used when referring to the full section view of these drawings as viewed from three feet above finished floor to show the locations of doorways, windows, stairwells, etc. Architectural drawings describe and document an architect's design.

Engineering

Engineering can be a very broad term. It stems from the Latin ingenerare, meaning "to create". Because this could apply to everything that humans create, it is given a narrower definition in the context of technical drawing. Engineering drawings generally deal with mechanical engineered items, such as manufactured parts and equipment.

Engineering drawings are usually created in accordance with standardized conventions for layout, nomenclature, interpretation, appearance (such as typefaces and line styles), size, etc.

Its purpose is to accurately and unambiguously capture all the geometric features of a product or a component. The end goal of an engineering drawing is to convey all the required information that will allow a manufacturer to produce that component.

SOFTWARE ENGINEERING

Software engineering practictioners make use of diagrams for designing software. Formal standards and modeling languages such as Unified Modeling Language (UML) exist but most diagramming happens using informal ad hoc diagrams that illustrate a conceptual model.

Practitioners reported that diagramming helped with analysing requirements, design, refactoring, documentation, onboarding, communication with stake holders. Diagrams are often transient or redrawn as required. Redrawn diagrams can acts as a form of shared understanding in a team.

6

EARTHQUAKE RESISTANT BUILDINGS

Falling under the category of natural disasters, the mere mention of earthquakes does not paint a very good picture. And rightfully so... the uprooting of families, the havoc within people, the loss of life, and the list goes on! However, in earlier times when the typology of structures was typically closer to the ground, not reaching a very substantial height, there would be scope to rush to open space for minimal damage. But in the present scenario with high-rise buildings jam-packed close to each other, there is left no option but to make the buildings strong enough to withstand the tectonic activities taking place in the region.

The field of Engineering and Architecture is fortunately on a constant roll in devising newer technologies and implementations day by day for earthquake-resistant buildings that take us a step closer to a safer tomorrow for our inhabitants.

There is a fundamental distinction between the design of a building and the construction methods used to fabricate that building. Advanced designs intended to withstand earthquakes are effective only if proper construction methods are used in the site selection, foundation, structural members, and connection joints. Earthquake-resistant designs typically incorporate ductility (the ability of a building to bend, sway, and deform without collapsing) within the structure and its structural members. A ductile building is able to bend and flex when exposed to the horizontal or vertical shear forces of an earthquake. Concrete buildings, which are normally brittle (relatively easy to break), can be made ductile by adding steel reinforcement. In buildings constructed with steel-reinforced concrete, both the steel and the concrete must be precisely manufactured to achieve the desired ductile behaviour.

Building failures during earthquakes often are due to poor construction methods or inadequate materials. In less-developed countries, concrete often is not properly mixed, consolidated, or cured to achieve its intended compressive strength, so buildings are thus extremely susceptible to failure under seismic loading. This problem is often made worse by a lack of local building codes or an absence of inspection and quality control.

Building failures are also frequently attributed to a shortage of suitable and locally available materials. For instance, when a building is designed with steel reinforced concrete, it is critical that the amount of steel used is not reduced to lower the building cost. Such practices substantially weaken a building's ability to withstand the dynamic forces of an earthquake.

Under normal conditions, a building's walls, columns, and beams primarily experience only vertical loads of compression. However, during an earthquake, lateral and shear loading occurs, which results in tensile and torsional forces on structural elements. Those forces result in high stresses at the building's corners and throughout various joints.

Strong construction joints are critical in building a structure that will withstand the shear loading of an earthquake. Since stress is concentrated at the joints between the walls, it is important that all the joints be properly prepared and reinforced. Concrete joints must also be properly compacted and anchored in order to achieve optimum strength. In the case of unreinforced masonry joints (mortar joints, such as those found in brick buildings), the anchoring between adjacent walls is especially important. When all the joints are tied together well, the building will act as a single integrated unit, enabling the forces of an earthquake to be transferred from one section to the next without catastrophic failure.

Earthquake-resistant construction requires that the building be properly grounded and connected through its foundation to the earth. Building on loose sands or clays is to be avoided, since those surfaces can cause excessive movement and nonuniform stresses to develop during an earthquake. Furthermore, if the foundation is too shallow, it will deteriorate, and the structure will be less able to withstand shaking. The foundation should therefore be constructed on firm soil to maintain a structure that settles uniformly under vertical loading.

Earthquake-resistant or aseismic structures are designed to protect buildings to some or greater extent from earthquakes. While no structure can be entirely immune to damage from earthquakes, the goal of earthquake-resistant construction is to erect structures that fare better during Seismic activity than their conventional counterparts. According to building codes, earthquake-resistant structures are intended to withstand the largest earthquake of a certain probability that is likely to occur at their location. This means the loss of life should be minimized by preventing collapse of the buildings for rare earthquakes while the loss of the functionality should be limited for more frequent ones.

To combat earthquake destruction, the only method available to ancient architects was to build their landmark structures to last, often by making them excessively stiff and strong.

Currently, there are several design philosophies in earthquake engineering, making use of experimental results, computer simulations and observations from past earthquakes to offer the required performance for the seismic threat at the site of interest. These range from appropriately sizing the structure to be strong and ductile enough to survive the shaking with an acceptable damage, to equipping it with base isolation or using structural vibration control technologies to minimize any forces and deformations. While the former is the method typically applied in most earthquake-resistant structures, important facilities, landmarks and cultural heritage buildings use the more advanced (and expensive) techniques of isolation or control to survive strong shaking with minimal damage. Examples of such applications are the Cathedral of Our Lady of the Angels and the Acropolis Museum.

TRENDS AND PROJECTS

Some of the new trends and/or projects in the field of earthquake engineering structures are presented.

Building materials

Based on studies in New Zealand, relating to Christchurch earthquakes, precast concrete designed and installed in accordance with modern codes performed well. According to the Earthquake Engineering Research Institute, precast panel buildings had good durability during the earthquake in Armenia, compared to precast frame-panels.

Earthquake shelter

One Japanese construction company has developed a six-foot cubical shelter, presented as an alternative to earthquake-proofing an entire building.

Concurrent shake-table testing

Concurrent shake-table testing of two or more building models is a vivid, persuasive and effective way to validate earthquake engineering solutions experimentally. Thus, two wooden houses built before adoption of the 1981 Japanese Building Code were moved to E-Defense for testing (see both pictures aside). The left house was reinforced to enhance its seismic resistance, while the other one was not. These two models were set on E-Defense platform and tested simultaneously.

COMBINED VIBRATION CONTROL SOLUTION

Designed by architect Merrill W. Baird of Glendale, working in collaboration with A. C. Martin Architects of Los Angeles, the Municipal Services Building at 633 East Broadway, Glendale was completed in 1966. Prominently sited at the corner of East Broadway and Glendale Avenue, this civic building serves as a heraldic element of Glendale's civic center.

In October 2004 Architectural Resources Group (ARG) was contracted by Nabih Youssef & Associates, Structural Engineers, to provide services regarding a historic resource assessment of the building due to a proposed seismic retrofit.

In 2008, the Municipal Services Building of the City of Glendale, California was seismically retrofitted using an innovative combined vibration control solution: the existing elevated building foundation of the building was put on high damping rubber bearings.

Steel plate walls system

A steel plate shear wall (SPSW) consists of steel infill plates bounded by a column-beam system. When such infill plates occupy each level within a framed bay of a structure, they constitute a SPSW system. Whereas most earthquake resistant construction methods are adapted from older systems, SPSW was invented entirely to withstand seismic activity.

SPSW behavior is analogous to a vertical plate girder cantilevered from its base. Similar to plate girders, the SPSW system optimizes component performance by taking advantage of the post-buckling behavior of the steel infill panels.

The Ritz-Carlton/JW Marriott hotel building, a part of the LA Live development in Los Angeles, California, is the first building in Los Angeles that uses an advanced steel plate shear wall system to resist the lateral loads of strong earthquakes and winds.

Kashiwazaki-Kariwa Nuclear Power Plant is partially upgraded

The Kashiwazaki-Kariwa Nuclear Power Plant, the largest nuclear generating station in the world by net electrical power rating, happened to be near the epicenter of the strongest Mw 6.6 July 2007 Chuetsu offshore earthquake. This initiated an extended shutdown for structural inspection which indicated that a greater earthquake-proofing was needed before operation could be resumed.

On May 9, 2009, one unit (Unit 7) was restarted, after the seismic upgrades. The test run had to continue for 50 days. The plant had been completely shut down for almost 22 months following the earthquake.

Seismic test of seven-story building

A destructive earthquake struck a lone, wooden condominium in Japan. The experiment was webcast live on July 14, 2009, to yield insight on how to make wooden structures stronger and better able to withstand major earthquakes.

The Miki shake at the Hyogo Earthquake Engineering Research Center is the capstone experiment of the four-year NEESWood project, which receives its primary support from the U.S. National Science Foundation Network for Earthquake Engineering Simulation (NEES) Program.

"NEESWood aims to develop a new seismic design philosophy that will provide the necessary mechanisms to safely increase the height of wood-frame structures in active seismic zones of the United States, as well as mitigate earthquake damage to low-rise wood-frame structures," said Rosowsky, Department of Civil Engineering at Texas A&M University. This philosophy is based on the application of seismic damping systems for wooden buildings. The systems, which can be installed inside the walls of most wooden buildings, include strong metal frame, bracing and dampers filled with viscous fluid.

SUPERFRAME EARTHQUAKE PROOF STRUCTURE

The proposed system is composed of core walls, hat beams incorporated into the top-level, outer columns, and viscous dampers vertically installed between the tips of the hat beams and the outer columns. During an earthquake, the hat beams and outer columns act as outriggers and reduce the overturning moment in the core, and the installed dampers also reduce the moment and the lateral deflection of the structure. This innovative system can eliminate inner beams and inner columns on each floor, and thereby provide buildings with column-free floor space even in highly seismic regions.

Earthquake Architecture

The term 'seismic architecture' or 'earthquake architecture' was first introduced in 1985 by Robert Reitherman. The phrase "earthquake architecture" is used to describe a degree of architectural expression of earthquake resistance or implication of architectural configuration, form or style in earthquake resistance. It is also used to describe buildings in which seismic design considerations impacted its architecture. It may be considered a new aesthetic approach in designing structures in seismic prone areas.

Earthquake-Resistant Materials

While shock absorbers, pendulums and "invisibility cloaks" may help dispel the energy to an extent, the materials chosen for a building are equally responsible for its stability.

Steel and Wood

For a material to resist stress and vibration, it must have high ductility, which is the ability to undergo large deformations and tension. Modern buildings are often constructed with structural steel, a component that comes in a variety of shapes and allows buildings to bend without breaking. Wood is also a surprising ductile material due to its high strength relative to its lightweight structure.

Innovative Materials

Scientists and engineers are developing new building materials with even greater shape retention. Innovations like shape memory alloys have the ability to both endure heavy strain and revert to their original shape. Additionally, fiber-reinforced plastic wrap — made by a variety of polymers — can be wrapped around columns and provide up to 38% added strength and ductility.

Engineers are also turning to natural elements to help reinforce buildings. The sticky yet rigid fibers of mussels and the strength-to-size ratio of spider silk have promising capabilities in creating structures. Bamboo and 3D printed materials can also function as lightweight, interlocking structures with limitless forms that can potentially provide even greater resistance for buildings.

Over the years, engineers and scientists have devised multiple techniques to create effective earthquake-proof buildings. However, as advanced as technology and materials are today, it is not always possible for buildings to completely withstand powerful earthquakes unscathed. Still, if a building is able to avoid collapse and save lives and communities, we can consider that a great success.

How Earthquakes Impact Buildings

Before we look at the features of earthquake-proof buildings, it's important to understand how earthquakes impact man-made structures. When an earthquake occurs, it sends shockwaves throughout the ground in short rapid intervals that extend in all directions. While buildings are generally equipped to handle vertical forces from their weight and gravity, they cannot traditionally handle side-to-side forces emitted by quakes.

This horizontal movement vibrates walls, floors, columns, beams and the connectors that hold them together. The difference in movement between the bottom and top of buildings exerts extreme stress, causing the supporting frame to rupture and the entire structure to eventually collapse.

Keys to Designing Earthquake-Resistant Buildings

When professionals design and construct buildings, they assess how to reduce risks. Following the applicable codes is one way to do that. Besides the international building codes that regulate the design, construction, alteration, and maintenance of new commercial and residential buildings, there are seismic codes. These are provisions that ensure structures can withstand earthquake forces.

Buildings made to withstand earthquakes may not look remarkable from the outside. However, numerous aspects make them more resilient during these disasters. Here are five of them:

1. An Appropriate Foundation

Creating a flexible foundation for a building could help it stay standing during an earthquake. One option is to build the structure on top of pads that separate the building from the ground. Then, the pads move, but the building stays still.

Another similar possibility, described in a 2019 research paper, is to place a solid foundation slab made of reinforced concrete and crisscrossing strips atop an intermediate cushion of sand.

This approach also included a trench around the foundation for further protection. Since this foundation design kept the building's base away from the soil, it was more resistant to seismic forces.

2. Seismic Dampers

Earthquake-resistant buildings also need features to help absorb shocks. People more commonly refer to them as seismic dampers. Engineers worked with NASA to develop damper systems for swing arms on its rockets in the 1960s. It chose a gas-driven shock isolation system first, then eventually progressed to a fluidics-based system that's still used today during space station launches and for earthquake-proofing buildings.

Seismic dampers absorb destructive energy, protecting the building from sustaining it. Generally, the larger the damper's diameter, the more force it can handle. One manufacturer of these dampers sells products to withstand from 25 to 1,100 tons and sells customized options, too.

Another approach involves putting a thin layer of graphene on top of a natural rubber pad. Researchers believe this will be a low-cost damper option for commercial and residential buildings.

3. A Drainage Mechanism

Pooled water can create structural complications. That's why parking garages often have double-tee load-bearing structures with a twist that lowers one corner — a feature called warping. Engineers achieve positive drainage with 1.5 percent minimum slopes across the diagonal toward floor drains. Drainage is also crucial to help structures tolerate earthquakes.

When the disasters occur in places with loose, sandy soils, the shaking can result in a phenomenon called liquefaction. It makes buildings sink or move to one side, and sewage pipes may rise to the surface. When the soil solidifies again after an earthquake, the buildings stay in their sunken, tilted positions.

However, earthquake drains help collected water escape, preventing liquefaction. They are prefabricated pieces wrapped in a filtering fabric. Each drain measures between 3 and 8 inches in diameter. A successful installation requires a grid-style placement. Depending on the size of the area prone to liquefaction, a building may need hundreds or thousands of drains.

4. Structural Reinforcement

Engineers and designers have various methods for strengthening a building's structure against potential earthquakes. Many of those redirect seismic forces. For example, shear walls and braced frames transfer lateral forces from the floors and roof to the foundation.

Then, diaphragms are rigid horizontal planes that move lateral forces to vertical-resistant parts of the building, such as a building's walls or framework. There are also movement-resistant frames. Those possibilities make a building frame's joints rigid while letting the other parts move.

Shorter buildings have less flexibility than taller ones. Thus, engineers typically realize they must provide more structural reinforcement for structures that are only a few stories tall versus skyscrapers.

5. Material With Adequate Ductility

Ductility describes how well a material can tolerate plastic deformation before it fails. Thus, materials with high ductility can absorb large amounts of energy without breaking. Structural steel is one of the most ductile materials, while brick and concrete are low-ductility materials.

Researchers have also developed creative solutions that show how structural steel is not the only earthquake-resistant material worth considering. For example, scientists engineered a fiber-reinforced concrete with properties similar to steel.

They called the material eco-friendly ductile cementitious composite. Experiments showed applying a 10-millimeter-thick layer to interior walls protected them from damage during a 9.0-magnitude simulated quake.

Projects are also underway to build earthquake-resistant residences in nations that lack the resources for safely built houses made from materials that people may need to import or lack the skills to use correctly — such as concrete and bricks. A civil engineering company showed how people in Indonesia could construct earthquake-resistant homes almost entirely from bamboo. The roofs feature corrugated sheets made from recycled Tetra Pak, a lightweight material that reflects heat.

How to Make A Building Earthquake-proof

To design an earthquake-proof building, engineers work to reinforce the structure and counteract a potential earthquake's forces. Since earthquakes release energy that pushes on buildings from one direction, the strategy involves having the building push the opposite way. Here are some of the methods used to help buildings withstand earthquakes.

1. Create a Flexible Foundation

One way to resist ground forces is to "lift" the building's foundation above the earth through a method called base isolation. Base isolation involves constructing a building on top of flexible pads made of steel, rubber and lead. When the base moves during an earthquake, the isolators vibrate while the structure itself remains steady. This effectively helps to absorb seismic waves and prevent them from traveling through the building.

2. Counter Forces with Damping

If you're familiar with shock absorbers used in cars, you might be surprised to learn that engineers also use a version of them in earthquake-resistant buildings. Similar to their use in cars, shock absorbers reduce the shockwaves' magnitude and help reduce pressure on the building. This is accomplished in two ways: vibrational control devices and pendulum power.

VIBRATIONAL CONTROL DEVICES

This method involves placing dampers at each level of a building between columns and beams. Each damper consists of piston heads inside a cylinder filled with silicone oil. When an earthquake occurs, the building transfers the vibrational energy into the pistons, which push against the oil. The energy is then transformed into heat, dissipating the force of the vibrations.

Pendulum Power

Another common damping method is pendulum power, used primarily in skyscrapers. To implement this, engineers suspend a large ball from steel cables that connect to a hydraulic system at the top of the building. When the building begins to sway, the ball acts as a pendulum and moves in the opposite direction to stabilize the building. Like damping, these features are tuned to match and counteract the building's movement in the event of an earthquake.

3. Shield Buildings from Vibrations

Rather than just counteracting forces, researchers are experimenting with ways buildings can deflect and reroute the energy from earthquakes altogether. Dubbed the "seismic invisibility cloak," this innovation involves creating a cloak of 100 concentric plastic and concrete rings and burying it at least 3 feet beneath the foundation of the building.

As seismic waves enter the rings, ease of travel forces them to move through to the outer rings. As a result, they are essentially channeled away from the building and dissipated into the ground.

4. Reinforce the Building's Structure

To withstand collapse, buildings must redistribute forces that travel through them during a seismic event. Shear walls, cross braces, diaphragms and moment-resisting frames are central to reinforcing a building.

Shear walls are a useful building technology that can help transfer earthquake forces. Made of multiple panels, these walls help a building keep its shape during movement. Shear walls are often supported by diagonal cross braces made of steel. These beams can support compression and tension, helping to counteract pressure and push forces.

Diaphragms are also a central part of a building's structure. Consisting of the building's floors, roof and the decks placed over them, diaphragms help remove tension from the floor and push forces to the building's vertical structures.

Moment-resisting frames provide additional flexibility in a building's design. These structures are placed among a building's joints and allow columns and beams to bend while the joints remain rigid. Thus, the building is able to resist the larger forces of an earthquake while still allowing designers the freedom to arrange building elements.

TECHNIQUES FOR EARTHQUAKE RESISTANT DESIGN OF STRUCTURES

There are many known and practiced measures to protect against seismic threats. Let's take a look at some of the earthquake resistant techniques used by the engineers world over to minimize the damage to structures due to earthquakes:

Floating Foundation:

The levitating or floating foundation separates the substructure of a building from its superstructure.

One way of doing this is by floating a building above its foundation on lead-rubber bearings that comprise a solid lead core covered in alternating layers of rubber and steel. The bearings are attached to the building and its foundation with the help of steel plates. So, when an earthquake occurs, the floating foundation can move without moving the structure above it.

In Japan this base isolation system works at a whole new level. Their design allows buildings to float mid-air. The system levitates, keeping the building on a cushion of air. The system has in-built sensors for detection of seismic activity and these sensors communicate with the air compressor that creates the layer of air between the building and its base.

Shock Absorption:

Similar to the shock absorbers used in vehicles, buildings also makes use of this technology. This earthquake resistant technology helps buildings slow down and reduce the magnitude of vibratory motions. Ideally shock absorbers should be placed at each level of the building – one end attached to the beam and the other end to the column. Each comprises a piston head that moves inside a cylinder full of silicone oil. During earthquakes, the horizontal motion of building will make the piston push against the oil, transforming mechanical energy from the quake to heat.

Rocking Core-Wall

Modern high-rise buildings use this technique to improve seismic resistance at a low cost. To make this work, a reinforced concrete core is set through the heart of the structure, surrounded by elevator banks. Many modern high-rise buildings use this technique to increase seismic resistance in an affordable way. It works most effectively when used together with base isolation. For base isolation, elastometric bearings are built with alternating layers of steel and natural rubber/neoprene.

The bearing thus created has low horizontal stiffness and vertical rigidity. The combination is highly effective, cost-friendly and simple to implement.

Pendulum Power:

The pendulum power technique works by suspending a huge mass near the top of the structure. This mass is supported by steel cables and viscous fluid dampers are placed between the mass and the building that it protects. In case of any seismic activity, the pendulum moves in the opposite direction to balance the energy. Each of the pendulums are tuned to sync with the natural frequency of the structure and these systems are called tuned mas dampers. Their goal is to counter resonance and reduce the structure's dynamic response.

Symmetry, Diaphragms And Cross-Bracing:

Generally one common criterion for seismic designs is symmetry. Seismic risks of asymmetrical designs are higher. L-Shaped, T-Shaped and split-level structures may be more visually appealing but they are also prone to torsion. Thus engineers design symmetrical structures to keep the forces equally distributed through the structure and limit ornamental elements like cornices, cantilever projections etc.

An earthquake has a significant lateral force. Seismic designing counteracts these forces in both horizontal and vertical structural systems. Diaphragms are integral to horizontal structures – such as floors of a building or roof. Engineers design each diaphragm on its own deck and strengthen it horizontally so it can distribute sideways forces with vertical structure parts.

With vertical structures, engineers have several approaches. Braced frames are often used in building walls. Braced frames rely on trusses for resisting sideways motion. Cross-bracing is a technique that uses two diagonal members in an X-shape to build wall trusses and it is a popular technique to build earthquake resistant structures.

EARTHQUAKE RESISTANT CONSTRUCTION TECHNIQUES

Natural disasters have adversely affected civilisations since ancient times. With advancements in technologies, we have learnt to mitigate its impacts on lives and properties.

Earthquakes are one of the most devastating forces on the planet. The seismic waves that travel through the ground can demolish buildings, kill people, and cost billions of dollars in damage and restoration. According to the National Earthquake Information Center, there are over 20,000 earthquakes every year on average, including 16 major disasters. The damage was caused by the collapse of buildings with people inside, as in previous earthquakes, prompting the development of earthquake-resistant constructions.

Constructions intended to withstand earthquakes are known as earthquake-resistant structures. While no structure can be completely safe from earthquake damage, earthquake-resistant construction aims to build structures that perform better than their conventional equivalents during seismic activity. Building rules state that earthquake-resistant constructions must be able to withstand the greatest earthquake with a reasonable chance of occurring at their site. There are now various design philosophies in earthquake engineering that use experimental results, computer models, and historical earthquake observations to provide the requisite performance for the seismic threat at the location of interest. In this article, we will deal with numerous techniques that can help improve a structure.

Earthquake is defined as a sudden ground shaking caused by the release of huge stored strain energy due to relative movements of the tectonic plates

Earthquake Resistant Construction is the fabrication of the building or a structure such that it may withstand the earthquakes.

Ground vibrations during earthquakes cause inertia forces at locations of mass in the structure. These forces travel through the roof /floors and walls to the foundation. We should ensure that these forces reach the ground without causing any major damage or collapse.

Characteristics features of Earthquake Resistant Construction Techniques

Helps in creating a flexible foundation to lift the building foundation above the earth.

It helps in shielding the building from the vibrations.

It also multiplies the safety strategies so that it can't collapse.

The materials which are used in earthquake-proof buildings minimise the chances of collapsing of buildings.

Pre Earthquake Resistant Construction Methods or Techniques

Basic techniques of the Earthquake Resistant Construction are as follows:

Base Isolation

- Energy dissipation devices
- Base Isolation devices

- These are the devices which are used to resist the earthquake to a greater extent.
- These devices help in separating the buildings from the building foundation by bearing pads.
- These are usually made from the Lead rubber Bearings, Spherical Sliding Isolation Systems.
- These are the devices which are supported by a series of bearing pads and these are placed between the building and building's foundation.
- The shaking is reduced by the 5 times due to this technique.
- In the case of an earthquake, the base-isolated building rocks back and forth like a boat to resist the seismic earthquakes.

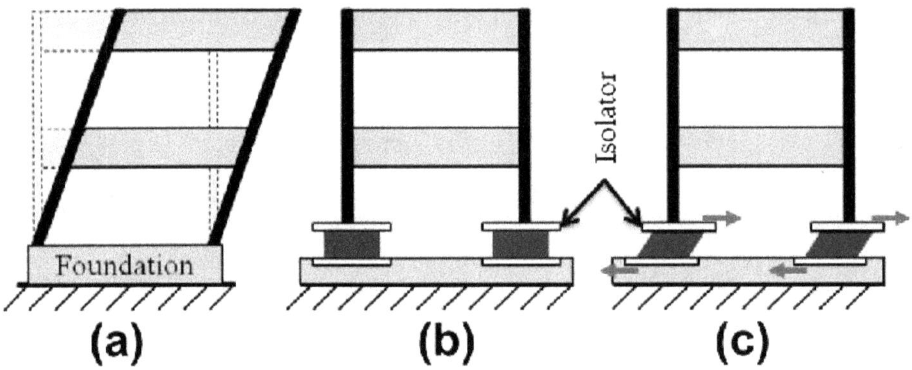

Fig 6.1

Energy dissipation devices

hese are the devices which are also termed as Seismic Dampers.

These are the special devices which are introduced in the building and they absorb the energy produced that is produced by the seismic waves.

They are of different types like the Viscous Dampers, Friction Dampers, and Metallic Dampers etc.

A viscous damper works effectively and utilizes the forced movement of the fluids within the damper.

A friction damper works efficiently and it utilizes the frictional energy to dissipate energy.

A metallic damper also utilizes the deformation of metal elements within the damper and helps in resisting seismic forces.

POST EARTHQUAKE RESISTANT CONSTRUCTION METHODS OR TECHNIQUES

Basic techniques of the Post Earthquake Resistant Construction are as follows:

In rural areas, there is a need for using better engineering information to achieve more durability.

The innovative concepts should be used like we should use the concrete blocks, cement, wooden blocks and the tiles to resist the earthquake up to a certain extent.

The construction should be according to the standard codes and techniques.

The structure should be properly reinforced.

The lives of the people should be least affected due to pre-earthquake resistant techniques.

The waste products should be utilized from the industries for the post-construction and the agricultural waste can be utilized in this context in the phase of the reconstruction.

Advantages in Long term of Earthquake Resistant Construction Techniques

The earthquake-resistant buildings can also tolerate the shocks and the vibrations which occur due to the seismic forces.

Earthquake Resistant Construction helps in reducing the number of structural damages.

- These techniques are helpful to resist other natural disasters.
- General Earthquake Resistant Construction Techniques
- The codes and the standards related to the earthquake-resistant construction should be followed to avoid this disaster up to a certain extent.
- We must avoid irregular shaped structures and framing system during the construction.
- We should use the best quality of construction materials.
- Proper workmanship must be ensured.
- We should also introduce the shear walls to transfer seismic loads to the bottom of the foundation so that they can resist the seismic forces.
- The Horizontal bands should be provided at plinth, lintel and roof levels as per IS codes
- The vertical reinforcement should be provided at important locations such as corners, internal and external wall junctions as per code.

The earthquakes do not kill the people but it is the unsafe structures which are responsible for the devastation. So, we should build earthquake resistant structures and keep societies safe.

Advanced Techniques

Advanced Techniques Some of the advanced techniques are:

Shock Absorption

Similar to the shock absorbers found in automobiles, structures employ this technology. This earthquake-resistant technology aids in the slowing down and reduction of the magnitude of vibrations in buildings. Shock absorbers should ideally be installed at each level of the structure, with one end attached to the beam and the other to the column. Each consists of a piston head that operates inside a silicone oil-filled cylinder. The horizontal motion of the building during an earthquake causes the piston to push against the oil, converting mechanical energy from the quake to heat.

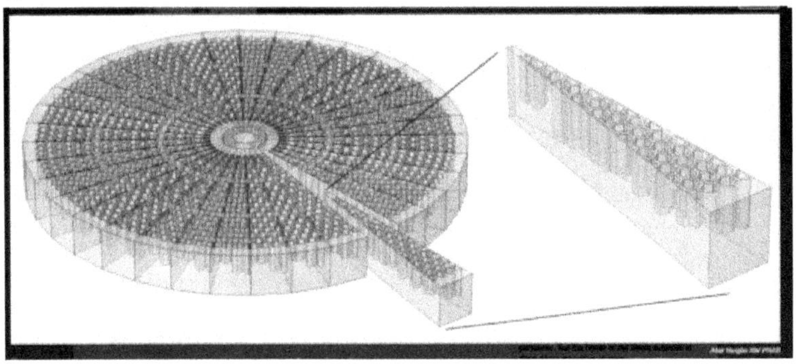

Fig 6.2

Flexible Foundation

Base isolation allow the building to be placed atop flexible pads that separate the foundation and allow the building to move in place. One technique to withstand ground stresses is to raise the foundation of the building above the ground. Here, in this case, the isolators vibrate as the base moves during an earthquake, but the structure itself stays stable. This effectively absorbs seismic vibrations and prevents them from passing through a structure.

Rocking-Core Wall –

Modern high-rise buildings use the rocking core-wall technology to improve seismic protection at a reasonable cost. A reinforced concrete core is established through the heart of the structure, encircled by elevator banks, to make this work.

This approach is used in many modern high-rise structures to boost seismic resilience in a cost-effective manner. When used in conjunction with base isolation, it is the most successful technique compared to other techniques. Elastometric bearings use alternating layers of steel and natural rubber/neoprene for base isolation. The resulting bearing has a low horizontal stiffness and a high vertical rigidity. The combination is highly effective, cost-efficient, and easy to use.

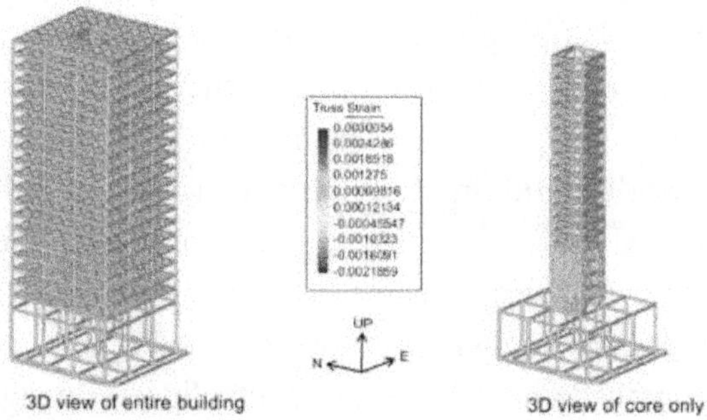

Fig 6.3

BUILDING STABILITY

Buildings must redistribute the forces that pass through them during a seismic event to avoid collapsing. Reinforcing a structure requires the use of shear walls, cross braces, diaphragms, and moment-resisting frames. In addition, shear walls are a valuable construction technique for transferring earthquake forces which is mentioned in IS 13920:2016. These walls, which are made of panels, allow a building maintain its shape during movement. Also, diagonal cross bracing are frequently used to support shear walls. These steel beams can support compression and tension, which helps to relieve pressure and redirect forces back to the foundation. Diaphragms are an important feature of the structure of a building as they help release tension from the floor and push force to the building's vertical structures. They are made up of the building's floors, roof, and decks placed over them. Moment-resisting frames allow for greater design flexibility while building a structure. This framework is installed between the building's joints and allows the columns and beams to flex while the joints remain stiff. As a result, the structure can withstand the greater forces of an earthquake while also giving designers more flexibility in how they arrange the building's elements.

Energy Dissipation Devices

This new technology for strengthening building relies on damping and energy dissipation as well, but it greatly extends the damping and energy dissipation given by lead-rubber bearings. Earthquake ground motion transfers a certain amount of vibration energy to the building. Buildings have the power to disperse, or dampen, this energy on their own. Buildings also have a finite ability to disperse energy before deformation and destruction occur.

This will release energy either through large-scale movement or increasing internal strains in structural parts such as columns and beams. Both of them eventually cause damage of differing degrees. As a result, by installing a building with additional high-damping-capacity devices, we may considerably reduce the seismic energy entering the structure and thereby reduce structural damage. As a result, a range of energy dissipation devices have been created and are now being used in real-world buildings. Damping devices are another name for energy dissipation devices. Many dampening devices that have been created can be classified into three categories: friction dampers: that dissipate energy by using frictional forces; metallic dampers: that rely on metal parts within the damper to deform; and viscoelastic dampers: that use controlled shearing of solids.

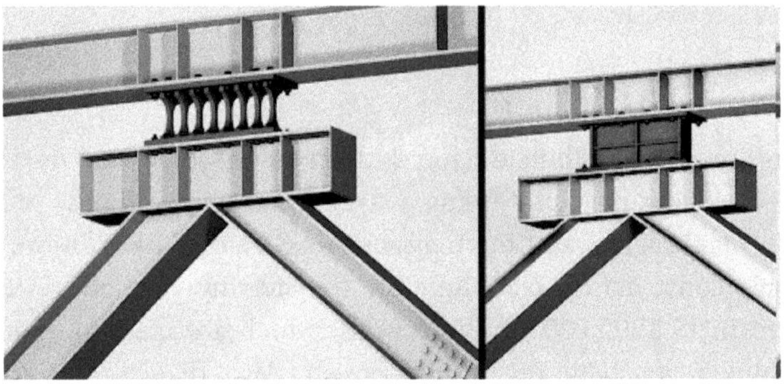

Fig 6.4

Pendulum Power –

This approach works by suspending a massive mass near the structure's top. Steel cables support the mass, and viscous fluid dampers are installed between the mass and the structure that it protects. In the event of seismic activity, the pendulum swings in the opposite direction to restore equilibrium. These systems are known as tuned mas dampers because each of the pendulums is tuned to match the structure's inherent frequency. Their purpose is to diminish the dynamic response of the structure by counteracting resonance.

Symmetry, Diaphragms, and Cross-Bracing –

Symmetry is a typical criterion used in seismic design. Asymmetrical designs have a higher seismic risk. Although L-shaped, T-shaped, and split-level constructions are more visually pleasing, they are also more susceptible to torsion.

As a result, engineers create symmetrical buildings to ensure that forces are distributed evenly throughout the structure and to keep aesthetic aspects such as cornices and cantilever projections to a minimum. A considerable lateral force is produced by an earthquake in both horizontal and vertical structural systems, and seismic design counteracts these stresses.

Diaphragms are essential components of horizontal constructions such as building floors and roofs. Engineers build each diaphragm on its own deck and strengthen it horizontally with vertical structural elements to distribute sideways forces. Also, braced frames are frequently utilised in the construction of walls in this sector and trusses are used to prevent sideways movement in braced frames.

Earthquake Generated Forces

To get a basic idea of how base isolation works, examine Figure 2. This shows an earthquake acting on both a base-isolated building and a conventional, fixed-base, building. As a result of an earthquake, the ground beneath each building begins to move. In Figure 2, it is shown moving to the left. Each building responds with movement which tends toward the right. The building undergoes displacement towards the right. The building's displacement in the direction opposite the ground motion is actually due to inertia. The inertial forces acting on a building are the most important of all those generated during an earthquake. It is important to know that the inertial forces which the building undergoes are proportional to the building's acceleration during ground motion. It is also important to realize that buildings don't actually shift in only one direction. Because of the complex nature of earthquake ground motion, the building actually tends to vibrate back and forth in varying directions. Base-Isolated, Fixed-Base Buildings

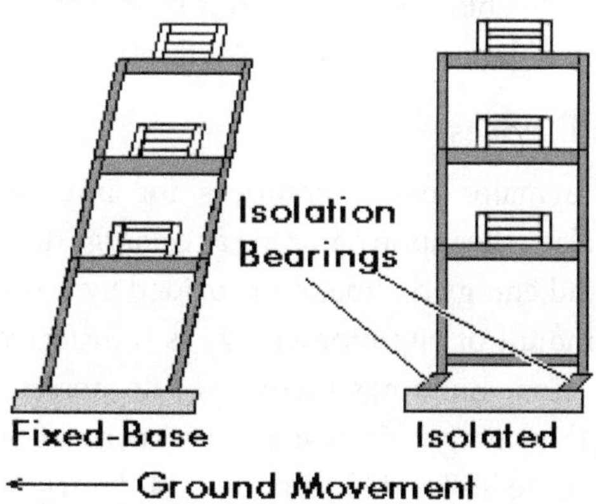

Fig 6.5

Deformation and Damages to Structures

In addition to displacing toward the right, the un-isolated building is also shown to be changing its shape-from a rectangle to a parallelogram. It is deforming. The primary cause of earthquake damage to buildings is the deformation which the building undergoes as a result of the inertial forces acting upon it.

RESPONSE OF BASE ISOLATED BUILDING

By contrast, even though it too is displacing, the base-isolated building retains its original, rectangular shape. It is the lead-rubber bearings supporting the building that are deformed. The base-isolated building itself escapes the deformation and damage, which implies that the inertial forces acting on the base-isolated building have been reduced. Experiments and observations of base-isolated buildings in earthquakes have been shown to reduce building accelerations to as little as 1/4 of the acceleration of comparable fixed-base buildings, which each building undergoes as a percentage of gravity. As we noted above, inertial forces increase, and decrease, proportionally as acceleration increases or decreases. Acceleration is decreased because the base isolation system lengthens a building's period of vibration, the time it takes for the building to rock back and forth and then back again. And in general, structures with longer periods of vibration tend to reduce acceleration, while those with shorter periods tend to increase or amplify acceleration. Finally, since they are highly elastic, the rubber isolation bearings don't suffer any damage. But the lead plug in the middle of our example bearing experiences the same deformation as the rubber. However, it generates heat. In other words, the lead plug reduces, or dissipates, the energy of motion, i.e., kinetic energy--by converting that energy into heat. And by reducing the energy entering the building, it helps to slow and eventually stop the building's vibrations sooner than would otherwise be the case, in other words, it damps the building's vibrations.

Energy Dissipation Devices

The second of the major new techniques for improving the earthquake resistance of buildings also relies upon damping and energy dissipation, but it greatly extends the damping and energy dissipation provided by lead-rubber bearings. As we've said, a certain amount of vibration energy is transferred to the building by earthquake ground motion. Buildings themselves do possess an inherent ability to dissipate, or damp, this energy. However, the capacity of buildings to dissipate energy before they begin to suffer deformation and damage is quite limited. The building will dissipate energy either by undergoing large scale movement or sustaining increased internal strains in elements such as the building's columns and beams. Both of these eventually result in varying degrees of damage. So, by equipping a building with additional devices which have high damping capacity, we can greatly decrease the seismic energy entering the building, and thus decrease building damage. Accordingly, a wide range of energy dissipation devices have been developed and are now being installed in real buildings. Energy dissipation devices are also often called damping devices. The large number of damping devices that have been developed can be grouped into three broad categories:

- **Friction Dampers:** these utilize frictional forces to dissipate energy
- **Metallic Dampers :** utilize the deformation of metal elements within the damper
- **Viscoelastic Dampers :** utilize the controlled shearing of solids
- **Viscous Dampers:** utilized the forced movement (orificing) of fluids within the damper

FLUID VISCOUS DAMPERS

General principles of damping devices are illustrated through Fluid Viscous damper. Following section, describes the basic characteristics of fluid viscous dampers, the process of developing and testing them, and the installation of fluid viscous dampers in an actual building to make it more earthquake resistant.

Damping Devices and Bracing Systems

Damping devices are usually installed as part of bracing systems. Figure 3 shows one type of damper-brace arrangement, with one end attached to a column and one end attached to a floor beam. Primarily, this arrangement provides the column with additional support. Most earthquake ground motion is in a horizontal direction; so, it is a building's columns which normally undergo the most displacement relative to the motion of the ground. Figure 3 also shows the damping device installed as part of the bracing system and gives some idea of its action.

7

SUBMISSION AND WORKING DRAWING

A working drawing is a drawing or blueprint based on explanations. It is completed with a thorough plan and views (details, notes, and dimensions) to ensure the product construction or replication without any additional information.

After learning what is working drawing in construction, you must be interested in knowing more about it. So, keep on reading.

Construction drawing is a set of drawings utilized for displaying the relation between varied parts of the project. In construction, the drawing will display all the axes of columns, all the dimensions including the ones needed to relate or assemble all the parts altogether or locations such as in between the walls, openings, and components. These are produced by the consultants.

They are also known as installation or execution drawings. These are used on the site to either build from scratch or see what has been built. These can be reviewed from time to time.

WORKING DRAWINGS

Working drawings, also known as good for construction drawings (GFC), are drawings provided by the architect that give detailed dimensioned, graphical information that can be used by a contractor to construct the works, or by suppliers to fabricate components of the works.

Production information is '...the information prepared by designers, which is passed to a construction team to enable a project to be constructed' (ref. CPIC The importance of production information). Production information is incorporated into tender documentation and then the contract documents for the construction works.

A working drawing is a part of production information provided by the designers to the construction team to help construct a project. Production information is of two types: working drawing and construction drawing.

Working drawings provide graphical, and dimensioned information. This data can be used by the contractors to build their work or help suppliers to create parts of the work or install/assemble the components. Working drawing can be in the form of structural drawings, architectural drawings, civil drawings, electrical drawings and mechanical drawings and so forth.

Conventionally, working drawings consist of two-dimensional orthogonal projections of the building including sections, plans, and elevations. These are either drawn by hand or using a Computer Aided Design (CAD) software. But now building information modeling (BIM) is used to make three-dimensional representations of buildings and the parts for construction.

Working drawings comprise of dimensions, title blocks, symbols, and notation. It is vital for them to be consistent with the industry standards to make an impact.

Creating a work of art is often an iterative process. This means that rather than diving feet-first into a complete drawing or painting, the artist will do a series of sketches with the aim of trying out or brainstorming ideas. It can be difficult to translate an idea from the mind to the canvas, so working drawings allow the artist to revise the work and re-draw to develop a composition, working through problems as they occur. Particularly in the case of large and complex works, these will then become references as the artist embarks on the final piece.

Fig 7.1

Working drawings are often among the most interesting of an artist's works because they reveal the thought processes behind a work of art; done not for an audience but for the artist's own use, they have an honesty and straightforwardness. As an artist, it's important not to let an awareness of that fact intrude on the function of your drawings. Particularly in the contemporary culture of documenting every moment, the intention to share work-in-progress on social media can lead to a sense of self-consciousness about the aesthetic of the drawing that can interfere with its primary roles of experimenting and inform the major work of art.

WORKING DRAWINGS IN DRAFTING AND ENGINEER-ING

Working drawings are drawings used as a reference or guide in the manufacture of a product. This most often refers to engineering and architecture, but working drawings are used in many different modes of construction. These drawings are composed according to industry standards so that all the information is easily and clearly understood, and standard conventions and units are used

There are two distinct types of working drawing: one is detail drawing, which shows various views of an object and includes important information such as measurements and tolerances that the craftsperson or machine operator might need to know when manufacturing the object, or that people using the object might need to know. The second is an assembly drawing, which shows how various components fit together during construction.

The Detail Drawing

The detail drawing conveys as much information as possible about a single component. It will be clearly labeled with a part number and name, It may include several views of the object--top, front and side--and a projection view. These drawings are annotated with information, including overall and detail dimensions, tolerances, materials, and treatments.

The Assembly Drawing

Assembly drawings show how pieces of a construction fit together. These may include an 'exploded' view, with pieces drawn separately but in correct relative positions, a 'general' assembly drawing where everything is drawn in its right place, and a detailed assembly drawing, which is a drawing of working assembly with measurements.

WORKING DRAWINGS IN ARCHITECTURE

Architectural working drawings need to not only show all the detail and measurement needed for the builder to construct the building but also to plan the construction process, particularly showing any unusual features or requirements that need special attention. These will include plans for each floor, exterior elevations (outside views) and sections (cutaway views) of the building.

Production information may include:

Drawings, such as working drawings.

Specifications.

Bills of quantities or schedules of work.

Working drawings provide dimensioned, graphical information that can be used; by a contractor to construct the works, or by suppliers to fabricate components of the works or to assemble or install components. They may include architectural drawings, structural drawings, civil drawings, mechanical drawings, electrical drawings, and so on.

Traditionally, working drawings consist of two-dimensional orthogonal projections of the building or component they are describing, such as plans, sections and elevations. These may be drawn to scale by hand, or prepared using Computer Aided Design (CAD) software.

However, increasingly, building information modelling (BIM) is being used to create three-dimensional representations of buildings and their components for construction. This may be described as a virtual construction model (VCM) and can comprise a number of different models prepared by different members of the project team.

Working drawings may include title blocks, dimensions, notation and symbols. It is important that these are consistent with industry standards so that their precise meaning is clear and can be understood. Specification information can be included on working drawings or in a separate specification, but information should not be duplicated as this can become contradictory and may cause confusion.

The scale at which drawings are prepared should reflect the level of detail of the information they are required to convey. Different line thicknesses can be used to provide greater clarity for certain elements.

It is important that the purpose of the drawings and the people that will use them are considered. Working drawings might be prepared for; statutory approvals, for contractors to plan the construction works, to provide instructions on site, for the procurement of components, for the preparation of shop drawings, for the appointment of subcontractors and so on.

Drawings must be structured carefully so that they convey necessary information to carry out particular parts of the works. To give greater clarity, they may be separated into packages, so that information is specifically tailored to separate parts of the works, specific components, or separate suppliers or trades.

It may be necessary to produce some packages earlier than others, for example, for items with long manufacturing times such as switchgear, chiller units, lifts, escalators or bespoke cladding systems, or for front-end construction such as service diversions, demolition, setting out details, underground drainage, piling and groundworks.

The quality of production information is extremely important. Unless it is prepared and co-ordinated properly, there will be disputes and delays on site, and costs will be incurred. Common problems with working drawings include:

- Poor co-ordination of information.
- Errors and omissions.
- Information not getting to the right people.
- Poor presentation.

Responsibility for the preparation of production information will depend on the selected system of procurement and the chosen form of contract. On traditional contracts (and management contracts and construction management contracts), production information may be produced by a consultant team, working for the client.

Some specialist elements of production information may be produced by specialist contractors, co-ordinated by the lead designer. On other forms of contract, such as design and build, responsibility for preparing and co-ordinating production information may lie with the main contractor.

Working drawings may be updated when the works are complete to show 'as constructed' information, reflecting changes to the works that may have occurred during the construction process.

Carefully prepared working drawings can be very beautiful and the very best have been exhibited as works of art.

NB Roles in construction projects: analysis and terminology, by Hughes, W. and Murdoch, J. R, published in 2001 by the University of Reading, suggests that working drawings is: 'A term that used to be common but seems to have fallen into disuse, describing information produced by designers for builders.'

Significance

These are one of the most important document released by the architect of a project because the accurate execution of these drawings determines the project's success. An important factor to consider when deciding on various working drawing components like dimensioning style, font, details, instructions, start points, north direction, key plan, etc. Is that they are intended for construction workers and clients. It should be chosen in such a way that it is self-explanatory, with as few errors as possible in its comprehension and execution.

Working drawings have traditionally been two-dimensional orthogonal projections of the building or components and are drafted to an appropriate legible scale which is easily readable. Today, Building information modeling (BIM) is increasingly being used to create three-dimensional representations of buildings and their components for construction. It also allows different stakeholders to collaborate easily as a virtual construction model (VCM), and can include a variety of models.

Typically Working Drawings fall into the Following Categories:

Architectural Drawings (It includes everything starting from plans, elevations, section drawings to the railing details, flooring, cladding, etc.)

Structural Drawings (These drawings comprise of details starting from foundation level to the plinth and then the slab)

HVAC, Electrical, Plumbing and Firefighting Drawings (Services are planned in advance by hiring consultants alongside for which drawings have to be produced in collaboration with them)

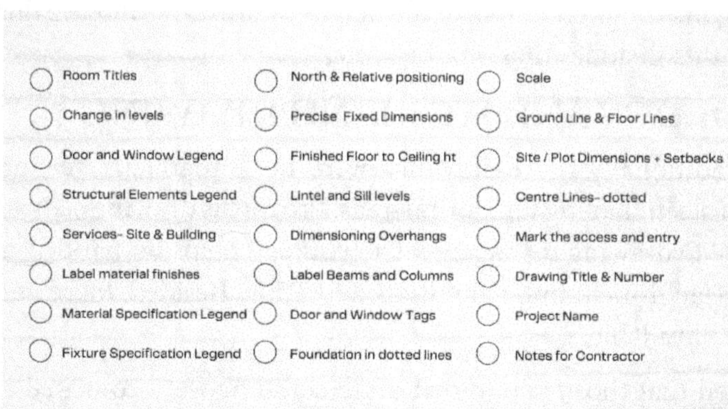

Fig 7.2

CHECKLIST FOR WORKING DRAWING DETAILS

Types of Working Drawing (In Construction)

What is working drawing in construction cannot be understood without knowing about its types. In construction, working drawing is further of 5 types. These include the following:

Architectural Drawing

An architectural drawing offers a complete view of a building. It exhibits the location of a building and all the parts of a building where they will be seen. Different architectural drawings are known by different names such as elevation, section, plan, and so on.

Structural Drawing

This kind of working drawing provides all the information regarding the structure such as the strength of various structural components, grade, size, structural materials, placement of reinforcement, etc.

Electrical Drawing

This type of working drawing offers details and the location of electrical wiring, sub-station, fixtures and so on. The electrical load calculation is also provided in the drawing.

Plumbing and Sanitary Drawings

This type of working drawings provides the location of piping for water supply, sanitary, fixture and every process that connects it to every fixture and so on.

Finishing Drawing

Finishing drawings comprise of the details of finishing and the final look of the building and this includes tiles, marbles, etc.

TECHNOLOGY AND WORKING DRAWING

While traditionally these drawings were done manually, technology is now playing a big role in shaping up the drawings. Procore is drawing centric software. As interactive CAD drawing management software, it helps in marking and attaching documents related including submittals and RFIs. Besides, there are many more CAD software available.

It is hoped that those who didn't understand or were confused about what is working drawing in construction, this article has helped them in understanding and clearing their doubts.

Essential Elements of a Set of Working Drawings

Whether you are building a custom home or building from a stock house plan, at the core of the project will be the working drawings. It is important to understand what working drawings are, what they are for, and what you need and don't need in the set of working drawings for your home.

Working drawings are integral to the construction process of your home. They are a set of drawings including all of the details and structural information required to both get a building permit and to build your home. When we put together a set of working drawings, we strive to prove what we call the "Minimum Optimum" set of plans. A lot of architectural firms will provide 30 pages of drawings when really only 10 would suffice. By paring this down to the essentials, it will save you money and confusion. The goal of your working drawings should be to have everything you need to build the home and nothing you don't. Let's take a look at what a set of working drawings from Mark Stewart Home Design will and won't have. We encourage you to click on each image to get a more detailed look at each of these plan pages.

Your Working Drawings Will Have:

Elevations

Elevations provide scaled views of the exterior of the home. This is where we include information about the roof slope, ceiling heights, exterior finishes, and more topographical information. The roof slope will be listed as a ratio of the vertical rise over a horizontal run. In the home pictured above, the roof has a slope of 4:12, or 4 inches of rise for every horizontal 12 inches. Crucial information about materials and techniques for finishing the exterior will also be given. Overhangs will be indicated where appropriate. These pages will also have additional details about what is required for the house plan to meet applicable building codes.

Floor Plans

The floor plans will include tons of information, and are shown as a horizontal cross section that will show all rooms, windows, and doors. Builders will be able to determine specifications for many materials, including framing components, cabinets and fixtures, doors, windows and much more. While some may choose to include separate framing sheets for floors and walls, this information is usually readily accessible via the floor plans. This is where the size of the individual rooms, location of windows, and location of things like fireplaces, appliances, kitchen islands, and plumbing fixtures will be shown. Floor plans will show all necessary information to make the sure the home meets fire life safety code (Currently the

2017 IRC). This is also the place where you will find information regarding exterior features like decks and garages.

Foundation Plan

The foundation plan is where you will find all of the critical information for the structural support features. The builder will be able to determine what type and what width of foundation wall is to be used, as well as locate and determine the size of the footings. Additional concrete features will be indicated where necessary. As you can see in the plan above, any additional information will given through very detailed notes regarding footings, point loads, construction techniques, and more.

Roof Framing Plan

These will provide accurate, detailed information about the structural members used to frame the roof. It will allow the builder to determine the number and size of the necessary structural members as well as the location. The roof framing plans will show information about the size and spacing of the rafters. Overhangs, venting and connectors will also be found here.

Section Views

Section views will provide very specific information regarding various construction and design features. Shown above, you will see a few examples representative of a detailed section view. You will notice larger cross section views of the home as well as smaller, more focused drawings of features like walls and stairs. Sometimes designers will display these sections as simple outline drawings and do not include key details. Remember, the function of a set of working drawings should be to provide everything needed to keep the building to code as well as everything needed to maintain the architectural integrity of the home. It is critical that these drawings show good structural detail and include detailed, specific notes.

Electrical Plans

Electrical plans will be included that show light fixtures, outlets, smoke and carbon monoxide detectors, switches and all code required fixtures. This Electrical plan will be a guid for you. Many people use it as a starting point and customize the layout to suit their lifestyle and particular requirements. Having built homes for years, and designed thousands of homes we are careful to give you what you need without adding unnecessary additional expense or complication.

CONSTRUCTION AND ARCHITECTURAL DETAILS

Our plan cross sections show much more detail then most. That being said sometimes additional construction and/or architectural details are necessary. Some plans need very few such details and some need pages of them. We use our best judgment from our dozens of years in the business as designers and builders to give you an optimum amount of detail for each particular house plan.

Your Working Drawings Won't Have:

Narrow Details or Minutiae

It is not necessary to include narrow and minute details about how to case each opening or finish each wall. A lot of these types of details will vary between individual builders and homeowners and are not included in the set of working drawings. While it is vital for the drawings to provide specific details, it is important to keep in mind whether or not these details are integral to the design and structural integrity of the home.

Fig 7.3

Detailed Plumbing Diagrams

In the floor plan we have shown above, you will see the accurate placement of the plumbing fixtures, but it is a misconception that you need to have detailed plumbing diagrams included in your set of working drawings. This was something that is sometimes needed in commercial buildings and was more commonly needed decades ago, but is not a requirement.

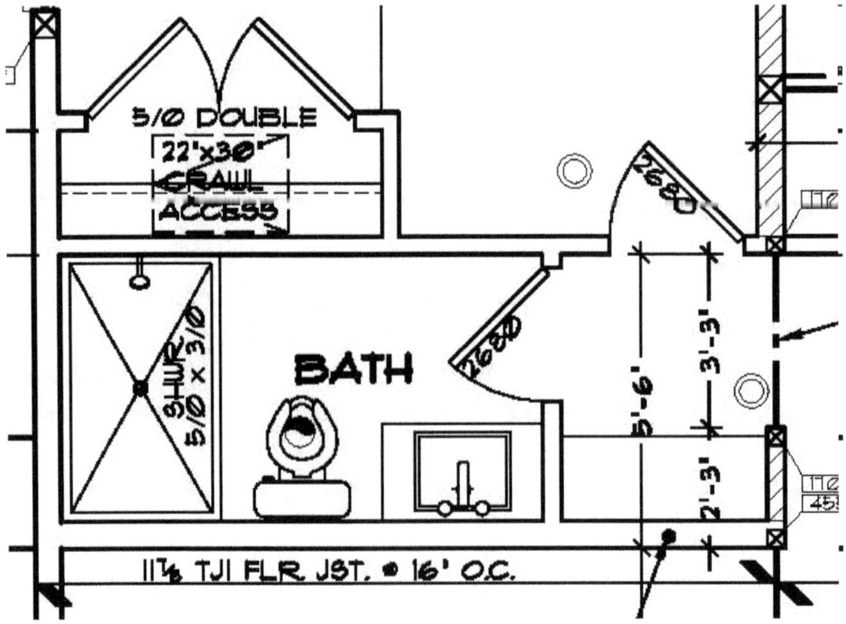

Fig 7.4

Site Plan (Only If you are having us prepare a custom design for you, or have otherwise contracted for it)

Site Plan Working Drawings

The site plan will include vital information about the lot that the home is to be built on. Your working drawings will provide information regarding easements, setbacks, property lines, and topography, as well as the footprint of the building, location of driveways, and any auxiliary buildings. The site plan also includes something called a benchmark, an elevation to which all other points of elevation will be referenced against. In the home shown above, it is marked as the top of an iron rod at 110 feet. Builders will use the benchmark as a starting point to calculate foundations and finished floor heights. Site plans also may include solar information that is crucial for orienting the home to maximize solar/energy efficiency. This is increasingly important for green building strategies and overall planning in conjunction with the rest of the house plan.

Submission Drawings

Any building construction cannot commence without due permission from the concerned local authority in charge. For getting a building permit, a set of drawings have to be submitted to the authority which are commonly called as submission drawings or permit drawings.

To understand the submission drawing, it is very important to first understand some of its basic requirements. Therefore, in this drawing, only some basic things have been explained to you. This submission drawing is very simple, which will give you a basic idea of how and on which the submission is actually made. If you understand this, then you can make a submission drawing of any building yourself.

A submission/Passing drawing must contains the following items;

- All Floor Plans
- Minimum one Section
- Minimum one elevation
- Site plan
- Key Plan
- North Direction
- Built up Area Calculations
- F.S.I. Area Calculations
- Parking Calculations
- Schedule of Openings
- Area Statement
- Project Title

Significance:

Permit Drawing must include precise information to determine whether the proposed work complies with all applicable regulations, the majority of which are concerned with life safety.

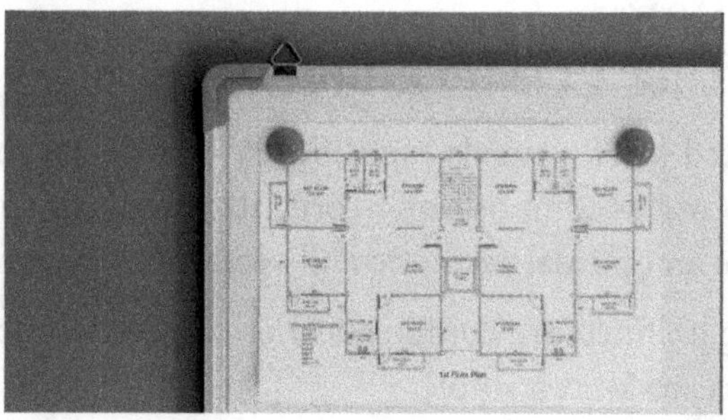

Fig 7.5

A substantially detailed sub-division plan, site plan, elevations, sections, all floor plans, and building services details are usually required. Space calculations are crucial in these drawings, a statement of FSI consumed, Total Built up area, Open Space, Amenity Space etc. should be clearly mentioned. To sum up, these drawings have more to do with the compliance of building regulations and the proposal.

Submission Drawings generally Include the Following:

1. **sub-division layout:** plan which shall be drawn to a scale 1:500

2. **Site Plan:** Should Mark the Position of the site, Means of access from the street, Setbacks to secure free circulation of air, admission of light and access. Existing physical features, such as wells, tanks, drains, pipelines, electrical provisions, trees, etc.

3. **Plans:** drawn to a scale of 1:100.

All floors together with the built-up area clearly indicating the sizes of rooms and the position and width of the staircase, ramps, and other exit ways, lift wells, lift machine room and lift pit details, meter room, and electric sub-station. Must include a ground floor plan and basement plan indicating the details of parking space. Dimensional details of doors, windows, and ventilators.

4. **Elevation and Section:** Sections showing the thickness of basement wall, wall construction, size and spacing of framing members, floors, slabs, roof slabs with the materials. Shall indicate the height of the building, rooms and parapet, drainage, and slope of the roof. (Mandatory section through the staircase) Dimensions of the projected portions like balconies and terrace plan indicating the drainage and the slope of the roof.

5. **Service plan** - Details of water supply, pump room, water tank, sewage disposal and treatment system, rainwater harvesting system and details of building services.

ARCHITECTURE COMPOSITION

Many art-related professionals, specifically architects and photographers , consider what is known as composition to organize physical elements using various principles to ensure an optimal aesthetic and visual arrangement.

When defining architecture composition, it can be described as "the planned arrangement of parts to form a whole.

Architects typically use it to compose concepts, contexts, drawings, elements, experiences, functions, materials, spaces, etc. Where the placement or order of elements such as lines, planes and volumes, or visual elements like color, texture, and size, form visual order and formal structure.

Why is composition important in architecture?

Architects use space and mass to form an ordered expression through composition. This ordering process is quintessential to direct a design's character, appearance, and style.

Composition is especially important in architecture for taking optimal photographs of a building or space, due to the relationship between the structural elements and negative, or surrounding, spaces.

Principles of Architecture composition

Proportion – Proportion as the relationship of one object to another or one part of an object to its other parts of the whole. Also, the proportion does not rely on a known size.

Scale – Scale refers to how the size of different elements relates to each other. Çankaya University suggests that varying the size of elements can produce visual drama that draws attention to a focal point.

The perception of scale can be affected by a viewer's own dimensions. Therefore, a viewer's dimensions play a role in the scale of the building he or she is looking at.

Repetition – Repetition is the implementation of the same colors, elements, etc. throughout the design. The principle can be used to create a pattern or repetition of an object or symbol in an art piece.

Rhythm – Rhythm is the repetitive use of one or more design elements to generate a sense of organized movement. This leads the eye to follow a visual link across the space.

Rhythm can be accomplished with repetition by duplicating shapes, colors, lines, and textures. One example would be the repetitive use of blue window panes of the same size on a building.

It can also be developed by a gradation of shifting size or color, via identical objects radiating from a center axis, through opposition by making abrupt changes in lines, shapes, and color. For example, alternating black and white floor tiles.

And by translation is utilizing curved lines to navigate your view across a straight surface. For instance, flowy curtains hanging along flat windows.

Contrast – Contrast is the difference between objects in a composition, usually from varying sizes, textures, colors, positions, shapes, orientation. This principle can create a feeling of depth within two or three-dimensional compositions.

While too much similarity between design elements can be dull, too much difference can cause confusion. Therefore, it is important to balance the level of contrast to develop a compelling visual.

Balance – Balance can be defined as the concept of visual equilibrium. Similar to our physical sense of balance, like attempting to stand on a ball, opposing forces within a composition generate visual stability.

Symmetrical balance, also known as formal balance, is achieved by having equal "weight" on each side of a central fulcrum. When the elements on each side are arranged identically then it is referred to as bilateral symmetry.

In a composition, balance can also be achieved by ordering elements equally around a central point. This method creates radial symmetry and is known as radial balance. In photography, the reflection of an object creates balance through inverted symmetry.

Asymmetrical balance, or informal balance, is the arrangement of elements of different visual weights to balance each other out along a central point. In architecture, this can take place by varying the design of each floor of a building.

In a composition, you can also create balance using color, texture, shapes, and spaces.

Restraint – Restraint is the measure or condition that keeps something under control.

In design, this means solving a problem in the simplest way possible. A good example of this principle is Minimalistic design, which is developed to support its main functionality without added ornamentation.

Architects who follow restraint may hold off on unnecessary details and exuberant colors to focus on material honesty and structural integrity. Mies van der Rohe's Farnsworth House is a famous example of using restrained colors and materials to develop a minimal space.

Definition – Definition references the outer line or form of any components or part of a building. Some examples include the exterior windows, archways, and different levels. In a composition, wide heavy columns can help define the edge of a building.

Hierarchy – Hierarchy is the placement of design elements into distinct levels of importance. This principle is used within design to emphasize one element over another and distinguish it as more important.

Therefore, visual hierarchy establishes prioritization and focal points.

Accentuation – Accentuation can be defined as the highlighting of certain parts of the architecture composition. To accomplish this, you must create a visual starting point for the eye to look at and navigate around the entire work.

Strength – In architectural composition, strength is the representation of durability and stability. This is achieved through the shape of a building; particularly a pyramid or rectangle. Also, stone or rough-textured materials, massive features, and structural elements help promote this principle.

Government buildings like the U.S. Capitol are extensively large and use columns to portray a feeling of strength.

Unity – Unity is when a harmony among the different design elements is achieved. More than that, it is the final result of each object and design principle working together to improve the appearance of a space.

Principles like balance, proportion, and rhythm within an area help bring the color, texture, and shape of objects into a practical relationship with the physical aspects of space, light, and structure.

8

PLANNING AND DESIGNING OF RESIDENTIAL BUILDING

Three complete sets of plans of your proposed residential construction project are required for the Building Section's plan check process. As well as communicating your construction project to the Building section and your contractor, a detailed set of plans helps you in your own planning, whether it be aesthetic of financial. In addition, a complete set of plans will help to identify and correct problems on paper, before construction begins, thus avoiding costly mistakes or changes during the construction process.

A complete set of working drawings will aid the Plans Examiner in assessing the structural integrity of your project so that your building is safe and comfortable. In addition, a Fire and Life Safety plan check is done to help assure that occupants have adequate exits, fire detection devices, etc., in case of a fire or other emergency.

In general, a complete, accurate set of drawings will help to speed up the plan check process and make any construction project less complex and frustrating.

Before planning a building or starting any construction, architects and planner must ensure the climatic condition of the particular location. Usually, local climate plays important role in the construction. Mainly, temperature (i.e. cold and hot), humidity, rain, snow, wind, and sunlight are the climatic factors that affect the building. All these notable factors play an important role in choosing the position of rooms, doors, windows, ventilation, balconies, and all other units of a building.

Apart from this, materials and techniques used for the construction highly depend upon the weather. Illustrating with it an example, in a country like India; majority regions experience a rainy season for the span of the 3-4 months, while the rest is winter and summer. This favours the houses which are built with brick, cement, concrete and also having a roof with terrace. Whereas, if we talk about the cold countries; there is snowfall for many months. Hence a roof without a terrace is preferred and at most wood is used as construction materials. Understanding this, we can sum up that the building's major features depend upon the climatic condition of the location.

Architectural engineering, also known as Building engineering, is the application of engineering principles and technology to building design and construction. Definitions of an architectural engineer may refer to:

- An engineer in the structural, mechanical, electrical, construction or other engineering fields of building design and construction.
- A licensed engineering professional in parts of the United States.
- In informal contexts, and formally in some places, a professional synonymous with or similar to an architect. In some languages, "architect" is literally translated as "architectural engineer".

The Process of Building a House

Acquire property

The first step in building a house is to look for suitable land, and/or an existing home that has potential to be reconstructed to meet your desires. There are many considerations that determine if the property is suitable. Some of these considerations include zoning, available utilities, topography restrictions, soil type, critical areas to avoid, the condition of existing structures, and local regulations. Typically, an architect, or civil engineer, can provide professional advice whether the land and/or structure is suitable for you.

find suitable land to build your home on

The location and type of land are important to keep in mind when developing a concept and when determining whether the concept is structurally viable.

Hire an architect

An architect will have knowledge of building code and design trends. Additionally, some states allow residential designers to practice residential design based on specific state law.

Architects will all have a different design aesthetic and preferences on construction types. Knowing your style preference should help you determine which architect fits your design criteria. It's also recommended to look at the architect's portfolio and ask for references and project examples. When debating on who to hire, consider your budget and how important trendy designs are to you. Be sure to ask other people involved in building a house (e.g. the builder) for their suggestions as well. But most importantly, look at the architect's portfolio of single-family homes.

Develop a concept

Work with your architect or home designer to develop a building concept and how it relates to the property. This includes everything from the architectural styles, building materials, site grading, and approximate construction costs.

Don't forget about the land, too. Think about why you chose it. Are there any aspects in particular that you like? Are neighborhood characteristics, school districts, or accessibility to the outdoors important to you? Consider these in your initial concept.

Architect or home designer collaborates with the engineer.

Collaboration between the architect and engineer is essential to develop an efficient structural envelope that addresses the overall desires of the client. In engineering speak, we call this "design development." Ideally, design development (DD) starts at early stages of project development such that structural construction types are considered early in the project.

Once the design development is vetted by the client, architect, and engineer the next push is for construction documents. These are your engineering plans for submittal to your local building department.

Seasonality and market conditions have a big impact on the building department review time. In our Pacific Northwest offices, there is usually a seasonal construction slow down. This typically translates into more favorable permit review times. In this case, it's recommended to start the project planning early, produce engineering construction documents in the winter, go in for building permit early spring, and start construction in early summer.

Engineering for building

Structural Engineering

Structural engineering involves the analysis and design of physical objects such as buildings, bridges, equipment supports, towers and walls. Those concentrating on buildings are responsible for the structural performance of a large part of the built environment and are, sometimes, informally referred to as "building engineers".

Structural engineers require expertise in strength of materials and in the seismic design of structures covered by earthquake engineering. Architectural Engineers sometimes practice structural as one aspect of their designs; the structural discipline when practiced as a specialty works closely with architects and other engineering specialists.

MECHANICAL, ELECTRICAL AND PLUMBING (MEP)

Mechanical and electrical engineers are specialists, commonly referred to as "MEP" (mechanical, electrical and plumbing) when engaged in the building design fields. Also known as "Building services engineering" in the United Kingdom, Canada and Australia. Mechanical engineers design and oversee the heating ventilation and air conditioning (HVAC), plumbing, and rain gutter systems. Plumbing designers often include design specifications for simple active fire protection systems, but for more complicated projects, fire protection engineers are often separately retained. Electrical engineers are responsible for the building's power distribution, telecommunication, fire alarm, signalization, lightning protection and control systems, as well as lighting systems.

The Architectural engineer (PE) in the United States

In many jurisdictions of the United States, the architectural engineer is a licensed engineering professional, usually a graduate of an architectural engineering university program preparing students to perform whole-building design in competition with architect-engineer teams; or for practice in one of structural, mechanical or electrical fields of building design, but with an appreciation of integrated architectural requirements. Formal architectural engineering education, following the engineering model of earlier disciplines, developed in the late 19th century, and became widespread in the United States by the mid-20th century. With the establishment of a specific "architectural engineering" NCEES Professional Engineering registration examination in the 1990s, and first offering in April 2003, architectural engineering became recognized as a distinct engineering discipline in the United States. Architectural engineers are not entitled to practice architecture unless they are also licensed as architects.

The Architect as Architectural Engineer

In some countries architecture, as a profession providing architectural services, is sometimes referred to as "architectural engineering". In others, such as in Japan, the terms "architecture" and "building engineering" are used synonymously. The practice of architecture includes the planning, designing and overseeing the building's construction.

In some languages, such as Korean and Arabic, "architect" is literally translated as "architectural engineer". In some countries, an "architectural engineer" (such as the ingegnere edile in Italy) is entitled to practice architecture and is often referred to as an architect. These individuals are often also structural engineers. In other countries, such as Germany, Austria and most of the Arabic countries, architecture graduates receive an engineering degree (Dipl.-Ing. - Diplom-Ingenieur).

In Brazil, architects and engineers currently share the same accreditation process (CREA - Regional Council of Engineers and Architects). Besides traditional architecture design training, Brazilian architecture courses also offer complementary training in engineering disciplines such as structural, electrical, hydraulic and mechanical engineering. After graduation, architects can be fully responsible for most engineering design and construction, except highly specialized tasks such as road design and high voltage electrical.

Education

The architectural, structural, mechanical and electrical engineering branches each have well established educational requirements that are usually fulfilled by completion of a university program.

Architectural Engineering as a single integrated field of study

What differentiates Architectural Engineering as a separate and single, integrated field of study, compared to other engineering disciplines, is its multi-disciplined engineering approach. Through training in and appreciation of architecture, the field seeks integration of building systems within its overall building design.

Architectural Engineering includes the design of building systems including Heating, ventilation and air conditioning (HVAC), plumbing, fire protection, electrical, lighting, Architectural acoustics, and structural systems. In some university programs, students are required to concentrate on one of the systems; in others, they can receive a generalist Architectural or Building Engineering degree.

HVAC (Heating, Ventilating, and Air Conditioning) refers to technology of indoor or automotive environmental comfort. HVAC system design is a major subdiscipline of mechanical engineering, based on the principles of thermodynamics, fluid mechanics, and heat transfer. Refrigeration is sometimes added to the field's abbreviation as HVAC&R or HVACR, or ventilating is dropped as in HACR (such as the designation of HACR-rated circuit breakers).

HVAC is important in the design of medium to large industrial and office buildings such as skyscrapers and in marine environments such as aquariums, where safe and healthy building conditions are regulated with temperature and humidity, as well as "fresh air" from outdoors.

Heating, ventilating, and air conditioning is based on inventions and discoveries made by Nikolay Lvov, Michael Faraday, Willis Carrier, Reuben Trane, James Joule, William Rankine, Sadi Carnot, and many others. The invention of the components of HVAC systems went hand-in-hand with the industrial revolution, and new methods of modernization, higher efficiency, and system control are constantly introduced by companies and inventors all over the world.

The three central functions of heating, ventilating, and air-conditioning are interrelated, providing thermal comfort, acceptable indoor air quality, within reasonable installation, operation, and maintenance costs. HVAC systems can provide ventilation, reduce air infiltration, and maintain pressure relationships between spaces. How air is delivered to, and removed from spaces is known as room air distribution.

In modern buildings the design, installation, and control systems of these functions are integrated into one or more HVAC systems. For very small buildings, contractors normally "size" and select HVAC systems and equipment. For larger buildings, building services designers and engineers, such as mechanical, architectural, or building services engineers analyze, design, and specify the HVAC systems, and specialty mechanical contractors build and commission them. Building permits and code-compliance inspections of the installations are normally required for all sizes of buildings.

The HVAC industry is a worldwide enterprise, with career opportunities including operation and maintenance, system design and construction, equipment manufacturing and sales, and in education and research. The HVAC industry had been historically regulated by the manufacturers of HVAC equipment, but Regulating and Standards organizations such as HARDI, ASHRAE, SMACNA, ACCA, Uniform Mechanical Code, International Mechanical Code, and AMCA have been established to support the industry and encourage high standards and achievement.

In planning a residential house, the following principles are to be kept in mind. Desh Pande has given certain principles which can be used as guides for residential planning.

These principles are:

1. Aspects:

The aspect of the house should be such that it enables the family members to live comfortably. Aspect is concerned with the orientation of the building. The arrangement of the doors and windows in the outside walls of the house should be in such a way that plenty of sunrays, breeze can enter into the house. All the rooms should be well ventilated and well lighted by the proper placement of doors and windows. Aspects of a building provide comfort, hygiene and pleasant views.

The kitchen should have an eastern aspect for getting pure air from outside which can remove the fumes of cooking and the room remains cool in the day time. Bed rooms should have a South East or South West aspect. The bed rooms should have a verandah in West. So that it protects the walls from getting direct sunrays. The drawing room or living room can have North East or South East aspect.

2. Prospect

Prospect is the view from outside of a house. The house should have a proper prospect so that it can give a feeling of cheerfulness to the people living in it. It should create a good impression on a person who views it from outside. Prospect should reveal pleasant features and hide unpleasant and undesirable features of the house.

The house not only looks attractive from outside, but it must also possess good qualities like comfort, cheerfulness, security, labour saving and modern outlook. For e.g. Small Projections or bay window can have a good out-look as well as helps in providing breeze, light and sunlight to the room.

3. Privacy

It is an important principle while planning a residential building.

Privacy can be maintained in ways:

(1) External

(2) Internal.

External Privacy:

This can be maintained by:

1. Fixing screens on the doors and windows.
2. If the house is built at the backside of the plot with enough distance from the highway and streets, then privacy can be maintained easily. The house should not be built at the edge of the roads.
3. Screens should be fixed on the main door of the house so that one cannot have a full view of the internal parts of the house from outside.
4. The external privacy can be maintained by planning the main entrance carefully or by screening it with trees or creepers.

Internal Privacy:

Internal privacy can be maintained by:

1. Screens may be used in the doors of all rooms which can easily separate one room from another. Bedrooms, bathrooms, toilet, kitchen should be properly screened. This can be possible if all rooms have independent access.
2. The door should be located on one side of the long wall of the room.
3. Privacy can be maintained by single panel doors.
4. Grouping: It means the rooms must be arranged in relation to one another. Different activity areas should be separated. The public or common rooms such as drawing room, living room, family room and dining room should be grouped together. The bedroom, study room, dressing room and bathroom should be separated from public areas by a central hall or passage. Dinning room should be close to the kitchen. The drawing room should not be near the kitchen. The bathrooms should be accessible to all rooms. Children's room should be near the parent's room.
5. Roominess: The roominess could be produced in a house by using small proportions. An impression of spaciousness can be created even if the house is small. A square room may look smaller than rectangular room of the same square meter area because of the change in proportion. The cupboards and shelves should be fixed in such a way as to make the best use of space available.

The built in wall cupboards, show cases, shelves help in avoiding over-crowding of floor area. The space under the window sill can be used for storage purposes. Wall cabinets in the kitchen provide enough storage space. The kitchen garden and the verandah should be so located that together they provide continuity of space.

6. Sanitation: The health of the family members depends upon the sanitary condition of the house. Provision should be made for proper lighting, ventilation, cleanliness and other sanitary conveniences in the house. All the rooms should be well lighted and well ventilated. There should be windows on opposite wall of the room to have cross ventilation. Instead of one window, there should be two to three windows in different walls of the room so that all the rooms can get sufficient light and air.

7. Flexibility: It means that the room can have multi-purpose use. When there is shortage of space, flexibility of room becomes important. For e.g. a living room

can be converted to bedroom at night. Kitchen should serve both the purposes of cooking and dinning. Bed rooms can be used as study room. Flexibility is important for those who live in flats.

8. Communication: Easy communication between rooms is one of the most important principles in residential house. This can be possible by the proper placement of doors, halls and stair cases. The doors should be placed in such a way that we can move without crossing the center of the room. The passage in the room should not interfere with the furniture arrangement.

A straight, short and separate passage is very important for privacy of every room. It helps in avoiding disturbances in doing work. Easy communication or circulation helps in reducing the movements of the members doing household activities.

9. Practical Consideration: Practical consideration plays an important role while constructing a house. The ease of cleaning and maintenance should be kept in mind while finishes are decided for floors and walls. Smooth and unbroken surfaces can be cleaned easily than the decorative pieces. The immediate surroundings of the house should be carefully considered. These are garden, courtyard or backyard. The garden around the house is important to beautify the house. The kitchen garden at the back of the house can be planned which can yield home grown vegetables and fruits.

The courtyard is the enclosed ground in front of the house where flower plants may be planted. Small tanks with water lilies or lotus plants and a fountain may be provided to make it attractive. The backyard is the enclosed space at the back of the house. This should be kept clean and may be used for some household activities. In planning a house, the needs of the family members must be taken into consideration.

DESIGN OF THE HVAC SYSTEM

The starting point in carrying out a heat estimate both for cooling and heating will depends on the ambient and inside conditions specified. However before taking up the heat load calculation, it is necessary to work out the fresh air requirement for each area in details, as pressurization is an important requirement.

Heating

There are many different types of standard heating systems. Central heating is often used in cold climates to heat private houses and public buildings. Such a system contains a boiler, furnace, or heat pump to heat water, steam, or air, all in a central location such as a furnace room in a home or a mechanical room in a large building. The use of water as the heat transfer medium is known as hydronics.

The system also contains either ductwork, for forced air systems, or piping to distribute a heated fluid and radiators to transfer this heat to the air. The term radiator in this context is misleading since most heat transfer from the heat exchanger is by convection, not radiation. The radiators may be mounted on walls or buried in the floor to give under-floor heat.

In boiler fed or radiant heating systems, all but the simplest systems have a pump to circulate the water and ensure an equal supply of heat to all the radiators. The heated water can also be fed through another (secondary) heat exchanger inside a storage cylinder to provide hot running water.

Forced air systems send heated air through ductwork. During warm weather the same ductwork can be used for air conditioning. The forced air can also be filtered or put through air cleaners.

Heating can also be provided from electric, or resistance heating using a filament that becomes hot when electric current is caused to pass through it. This type of heat can be found in electric baseboard heaters, portable electric heaters, and as backup or supplemental heating for heat pump (or reverse heating) system.

The heating elements (radiators or vents) should be located in the coldest part of the room, typically next to the windows to minimize condensation and offset the convective air current formed in the room due to the air next to the window becoming negatively buoyant due to the cold glass.

Devices that direct vents away from windows to prevent "wasted" heat defeat this design intent. Cold air drafts can contribute significantly to subjectively feeling colder than the average room temperature. Therefore, it is important to control the air leaks from outside in addition to proper design of the heating system.

The invention of central heating is often credited to the ancient Romans, who installed a system of air ducts called a hypocaust in the walls and floors of public baths and private villas.

Ventilating

Ventilating is the process of "changing" or replacing air in any space to control temperature or remove moisture, odors, smoke, heat, dust, airborne bacteria, carbon dioxide, and to replenish oxygen. Ventilation includes both the exchange of air to the outside as well as circulation of air within the building. It is one of the most important factors for maintaining acceptable indoor air quality in buildings.

Methods for ventilating a building may be divided into mechanical/forced and natural types. Ventilation is used to remove unpleasant smells and excessive moisture, introduce outside air, to keep interior building air circulating, and to prevent stagnation of the interior air.

Mechanical or forced ventilation

"Mechanical" or "forced" ventilation is provided by an air handler and used to control indoor air quality. Excess humidity, odors, and contaminants can often be controlled via dilution or replacement with outside air. However, in humid climates much energy is required to remove excess moisture from ventilation air.

Kitchens and bathrooms typically have mechanical exhaust to control odors and sometimes humidity. Factors in the design of such systems include the flow rate (which is a function of the fan speed and exhaust vent size) and noise level. If ducting for the fans traverse unheated space (e.g., an attic), the ducting should be insulated as well to prevent condensation on the ducting. Direct drive fans are available for many applications, and can reduce maintenance needs.

Ceiling fans and table/floor fans circulate air within a room for the purpose of reducing the perceived temperature because of evaporation of perspiration on the skin of the occupants. Because hot air rises, ceiling fans may be used to keep a room warmer in the winter by circulating the warm stratified air from the ceiling to the floor. Ceiling fans do not provide ventilation as defined as the introduction of outside air.

NATURAL VENTILATION

Natural ventilation is the ventilation of a building with outside air without the use of a fan or other mechanical system. It can be achieved with openable windows or trickle vents when the spaces to ventilate are small and the architecture permits. In more complex systems warm air in the building can be allowed to rise and flow out upper openings to the outside (stack effect) thus forcing cool outside air to be drawn into the building naturally through openings in the lower areas.

These systems use very little energy but care must be taken to ensure the occupants' comfort. In warm or humid months, in many climates, maintaining thermal comfort solely via natural ventilation may not be possible so conventional air conditioning systems are used as backups. Air-side economizers perform the same function as natural ventilation, but use mechanical systems' fans, ducts, dampers, and control systems to introduce and distribute cool outdoor air when appropriate.

Air conditioning

Air conditioning and refrigeration are provided through the removal of heat. The definition of cold is the absence of heat and all air conditioning systems work on this basic principle. Heat can be removed through the process of radiation, convection, and Heat cooling through a process called the refrigeration cycle. The conduction mediums such as water, air, ice, and chemicals are referred to as refrigerants.

An air conditioning system, or a standalone air conditioner, provides cooling, ventilation, and humidity control for all or part of a house or building.

The refrigerant cycle consists of four essential elements to create a cooling effect. The system refrigerant starts its cycle in a gaseous state. The compressor pumps the refrigerant gas up to a high pressure and temperature. From there it enters a heat exchanger (sometimes called a "condensing coil") where it loses energy (heat) to the outside. In the process the refrigerant condenses into a liquid.

The liquid refrigerant is returned indoors to another heat exchanger ("evaporating coil"). A metering device allows the liquid to flow in at a low pressure at the proper rate. As the liquid refrigerant evaporates it aborbs energy (heat) from the inside air, returns to the compressor, and the cycle repeats. In the process, heat is absorbed from indoors, and transferred outdoors, resulting in cooling of the building.

Central, 'all-air' air conditioning systems are often installed in modern residences, offices, and public buildings, but are difficult to retrofit (install in a building that was not designed to receive it) because of the bulky air ducts required. A duct system must be carefully maintained to prevent the growth of pathogenic bacteria in the ducts.

An alternative to large ducts to carry the needed air to heat or cool an area is the use of remote fan coils or split systems. These systems, although most often seen in residential applications, are gaining popularity in small commercial buildings. The evaporator coil is connected to a remote condenser unit using piping instead of ducts.

Dehumidification in an air conditioning system is provided by the evaporator. Since the evaporator operates at a temperature below dew point, moisture in the air condenses on the evaporator coil tubes. This moisture is collected at the bottom of the evaporator in a condensate pan and is removed by piping it to a central drain or onto the ground outside.

A dehumidifier is an air-conditioner-like device that controls the humidity of a room or building. It is often employed in basements which have a higher relative humidity because of their lower temperature (and propensity for damp floors and walls). In food retailing establishments, large open chiller cabinets are highly effective at dehumidifying the internal air. Conversely, a humidifier increases the humidity of a building.

Air-conditioned buildings often have sealed windows, because open windows would disrupt the attempts of the HVAC system to maintain constant indoor air conditions. All modern air conditioning systems, down to small "window" units, are equipped with internal air filters. These are generally of a light weight gauze-type element, and must be replaced as conditions warrant (some models may be washable). For example, a building in a high-dust environment, or a home with furry pets, will need to have the filters changed more often than buildings without these dirt loads.

Failure to replace these filters as needed will contribute to a lower heat-exchange rate, resulting in wasted energy, shortened equipment life, and higher energy bills; also low air flow can result in "iced-up" or "iced-over" evaporator coils, and then there is no air flow at all. Additionally, very dirty or plugged filters can cause overheating during a heating cycle, and can possibly result in damage to the furnace unit or even fire.

It is important to keep in mind that because an air conditioner moves heat from the indoor (evaporator) coil to the outdoor (condenser) coil, the latter must be kept just as clean as the former. This means that, in addition to replacing the air filter at the evaporator coil, it is also necessary to regularly clean the condenser coil.

Failure to keep the condenser clean will eventually result in harm to the compressor, because the condenser coil is responsible for discharging both the indoor heat (as picked up by the evaporator) plus the heat generated by the electric motor driving the compressor.

Outside, "fresh" air is generally drawn into the system by a vent into the evaporator section. Adjustment of the percentage of return air made up of fresh air can usually be adjusted by manipulating the opening of this vent.

ENERGY EFFICIENCY

For the last 20 to 30 years, manufacturers of HVAC equipment have been making an effort to make the systems they manufacture more efficient. This was originally driven by rising energy costs, and has more recently been driven by increased awareness of environmental issues. In the USA, the EPA has also imposed tighter restrictions. There are several methods for making HVAC systems more efficient.

Heating energy

Water heating is more efficient for heating buildings and was the standard many years ago. Today forced air systems can double for air conditioning and are more popular.

A couple of benefits of forced air systems, which are now widely applied in churches, schools and high-end residences,are 1) better air conditioned effect 2) up to 15-20% energy saving, and 3) evenly conditioned effect. A drawback is the installation cost, which might be slightly higher than traditional HVAC system.

Energy efficiency can be improved even more in central heating systems by introducing zoned heating. This allows a more granular application of heat, similar to non-central heating systems. Zones are controlled by multiple thermostats. In water heating systems the thermostats control zone valves, and in forced air systems they control zone dampers inside the vents which selectively block the flow of air. In this case, the control system is very critical to maintain a proper temperature.

GEOTHERMAL HEAT PUMP

Geothermal heat pumps are similar to ordinary heat pumps, but instead of using heat found in outside air, they rely on the stable, even heat of the earth to provide heating, air conditioning and, in most cases, hot water. From Montana's ?70 °F (?57 °C) temperature, to the highest temperature ever recorded in the U.S.-134 °F (56.7 °C) in Death Valley, California, in 1913-many parts of the country experience seasonal temperature extremes.

A few feet below the earth's surface, however, the ground remains at a relatively constant temperature. Although the temperatures vary according to latitude, at 6 feet (1.83 m) underground, temperatures range from 45 to 75 °F (7.2 to 23.9 °C).

While they may be more costly to install initially than regular heat pumps, they can produce markedly lower energy bills-30 percent to 40 percent lower, according to estimates from the U.S. Environmental Protection Agency.

Ventilation energy recovery

Energy recovery systems sometimes utilize heat recovery ventilation or energy recovery ventilation systems that employ heat exchangers or enthalpy wheels to recover sensible or latent heat from exhausted air. This is done by transfer of energy to the incoming outside fresh air.

Air conditioning energy

The performance of vapor compression refrigeration cycles is limited by thermodynamics. These air conditioning and heat pump devices move heat rather than convert it from one form to another, so thermal efficiencies do not appropriately describe the performance of these devices. The Coefficient-of-Performance (COP) measures performance, but this dimensionless measure has not been adopted, but rather the Energy Efficiency Ratio (EER). EER is the Energy Efficiency Ratio based on a 35 °C (95 °F) outdoor temperature.

To more accurately describe the performance of air conditioning equipment over a typical cooling season a modified version of the EER is used, and is the Seasonal Energy Efficiency Ratio (SEER). SEER ratings are based on seasonal temperature averages instead of a constant 35 °C outdoor temperature. The current industry minimum SEER rating is 13 SEER. The SEER article describes it further, and presents some economic comparisons using this useful performance measure.

Engineers have pointed out some areas where efficiency of the existing hardware could be improved. For example, the fan blades used to move the air are usually stamped from sheet metal, an economical method of manufacture, but as a result they are not aerodynamically efficient. A well-designed blade could reduce electrical power required to move the air by a third.

A central heating system provides warmth to the whole interior of a building (or portion of a building) from one point to multiple rooms. When combined with other systems in order to control the building climate, the whole system may be a HVAC (heating, ventilation and air conditioning) system.

Central heating differs from local heating in that the heat generation occurs in one place, such as a furnace room in a house or a mechanical room in a large building (though not necessarily at the "central" geometric point). The most common method of heat generation involves the combustion of fossil fuel in a furnace or boiler. The resultant heat then gets distributed: typically by forced-air through ductwork, by water circulating through pipes, or by steam fed through pipes. Increasingly, buildings utilize solar-powered heat sources, in which case the distribution system normally uses water circulation.

In much of northern Europe and in urban portions of Russia, where people seldom require air conditioning in homes due to the temperate climate, most new housing comes with central heating installed. Such areas normally use gas heaters, district heating, or oil-fired systems. In the western and southern United States natural-gas-fired central forced-air systems occur most commonly; these systems and central-boiler systems both occur in the far northern regions of the USA.

Steam-heating systems, fired by coal, oil or gas, feature in the USA, Russia and Europe: primarily for larger buildings. Electrical heating systems occur less commonly and are only practical with low cost electricity or when geothermal heat pumps are used. Considering the combined system of central generating plant and electric resistance heating, the overall efficiency will be less than for direct use of fossil fuel for space heating.

Some buildings in the Roman Empire used central heating systems, conducting air heated by furnaces through empty spaces under the floors and out of pipes in the walls - a system known as a hypocaust. A similar system of central heating was used in ancient Korea, where it is known as ondol. It is thought that the ondol system dates back to the Koguryo or Three Kingdoms (37 BC-AD 668) period when excess heat from stoves were used to warm homes.

In the early medieval Alpine upland, a simpler central heating system where heat travelled through underfloor channels from the furnace room replaced the Roman hypocaust at some places. In Reichenau Abbey a network of interconnected underfloor channels heated the 300 m^2 large assembly room of the monks during the winter months. The degree of efficiency of the system has been calculated at 90%.

In the 13th century, the Cistercian monks revived central heating in Christian Europe using river diversions combined with indoor wood-fired furnaces. The well-preserved Royal Monastery of Our Lady of the Wheel (founded 1202) on the Ebro River in the Aragon region of Spain provides an excellent example of such an application.

The Roman hypocaust continued to be used on a smaller scale during late Antiquity and by the Umayyad caliphate, while later Muslim builder employed a simpler system of underfloor pipes.

By about 1700 Russian engineers had started designing hydrologically based systems for central heating. The Summer Palace (1710-1714) of Peter the Great in Saint Petersburg provides the best extant example. Slightly later, in 1716, came the first use of water in Sweden to distribute heat in buildings. Martin Triewald, a Swedish engineer, used this method for a greenhouse at Newcastle upon Tyne. Jean Simon Bonnemain (1743-1830), a French architect, introduced the technique to industry on a cooperative, at Château du Pêcq, near Paris.

Angier March Perkins developed and installed some of the earliest steam-heating systems in the 1830s. The first was installed in the home of Governor of the Bank of England John Horley Palmer so that he could grow grapes in England's cold climate.

Franz San Galli, a Polish-born Russian businessman living in St. Petersburg, invented the radiator between 1855-1857, which was a major step in the final shaping of modern central heating.

WATER HEATING

Common components of a central heating system using water-circulation include:

- Gas supply lines (sometimes including a propane tank), oil tank and supply lines or district heating supply lines
- Boiler (or a heat exchanger for district heating) - heats water in a closed-water system
- Pump - circulates the water in the closed system
- Radiators - wall-mounted panels through which the heated water passes in order to release heat into rooms

Engineers in the United Kingdom and in other parts of Europe commonly combine the needs of room heating with hot-water heating and storage. These systems occur less commonly in the USA. In this case, the heated water in a sealed system flows through a heat exchanger in a hot-water tank or hot-water cylinder where it heats water from the normal water supply before that water gets fed to hot-water outlets in the house. These outlets may service hot-water taps or appliances such as washing machines or dishwashers.

Sealed water-circulating system

A sealed system provides a form of central heating in which the water used for heating usually circulates independently of the building's normal water supply. An expansion tank contains compressed gas, separated from the sealed-system water by a diaphragm. This allows for normal variations of pressure in the system.

A safety valve allows water to escape from the system when pressure becomes too high, and a valve can open to replenish water from the normal water supply if the pressure drops too low. Sealed systems offer an alternative to open-vent systems, in which steam can escape from the system, and gets replaced from the building's water supply via a feed and central storage system.

ELECTRIC AND GAS-FIRED HEATERS

Electric heating or resistance heating converts electricity directly to heat. Electric heat is often more expensive than heat produced by combustion appliances like natural gas, propane, and oil. Electric resistance heat can be provided by baseboard heaters, space heaters, radiant heaters, furnaces, wall heaters, or thermal storage systems.

Electric heaters are usually part of a fan coil which is part of a central air conditioner. They circulate heat by blowing air across the heating element which is supplied to the furnace through return air ducts. Blowers in electric furnaces move air over one to five resistance coils or elements which are usually rated at five kilowatts. The heating elements activate one at a time to avoid overloading the electrical system. Overheating is prevented by a safety switch called a limit controller or limit switch. This limit controller may shut the furnace off if the blower fails or if something is blocking the air flow. The heated air is then sent back through the home through supply ducts.

In larger commercial applications, central heating is provided through an air handler which incorporates similar components as a furnace but on a larger scale.

Hydronic and steam systems

Hydronic heating systems are systems that circulate a medium for heating. Hydronic radiant floor heating systems use a boiler or district heating to heat water and a pump to circulate the hot water in plastic pipes installed in a concrete slab. The pipes, embedded in the floor, carry heated water that conducts warmth to the surface of the floor where it broadcasts heat energy to the room above.

Hydronic systems circulate hot water for heating. Steam heating systems are similar to heating water systems, except steam is used as the heating medium instead of water.

Hydronic heating systems generally consist of a boiler or district heating heat exchanger, hot water circulating pumps, distribution piping, and a fan coil unit or a radiator located in the room or space. Steam heating systems are similar except no circulating pumps are required.

Hydronic systems are closed loop: the same fluid is heated and then reheated. Hydronic heating systems are also used with antifreeze solutions in ice and snow melt systems for walkways, parking lots and streets. They are more commonly used in commercial and whole house radiant floor heat projects, while electric radiant heat systems are more commonly used in smaller "spot warming" applications.

Heat pumps

In mild climates a heat pump can be used to air condition the building during hot weather, and to warm the building using heat extracted from outdoor air in cold weather. Air-source heat pumps are generally uneconomic for outdoor temperatures much below freezing. In colder climates, geothermal heat pumps can be used to extract heat from the ground. For economy, these systems are designed for average low winter temperatures and use supplemental heating for extreme low temperature conditions.

The advantage of the heat pump is that it reduces the purchased energy required for building heating; often geothermal source systems also supply domestic hot water. Even in places where fossil fuels provide most electricity, a geothermal system may offset greenhouse gas production since most of the energy furnished for heating is supplied from the environment, with only 15-30% purchased.

Environmental aspects

From an energy-efficiency standpoint considerable heat gets lost or goes to waste if only a single room needs heating, since central heating has distribution losses and (in the case of forced-air systems particularly) may heat some unoccupied rooms without need. In such buildings which require isolated heating, one may wish to consider non-central systems such as individual room heaters, fireplaces or other devices. Alternatively, architects can design new buildings to use low-energy building techniques which can virtually eliminate the need for heating, such as those built to the Passive House standard.

However, if a building does need full heating, combustion central heating offers a more environmentally friendly solution than electric-air central heating or than other direct electric heating devices. This stems from the fact that most electricity originates remotely using fossil fuels, with up to two-thirds of the energy in the fuel lost (unless utilized for district heating) at the power station and in transmission losses. In Sweden proposals exist to phase out direct electric heating for this reason. Nuclear and hydroelectric sources reduce this factor.

In contrast, hot-water central heating systems can use water heated in or close to the building using high-efficiency condensing boilers, biofuels, or district heating. Wet underfloor heating has proven ideal. This offers the option of relatively easy conversion in the future to use developing technologies such as heat pumps and solar combisystems, thereby also providing future-proofing.Typical efficiencies for central heating are: 85-97% for gas fired heating; 80-89% for oil-fired, and 45-60% for coal-fired heating.

Piping

Within industry, piping is a system of pipes used to convey fluids (liquids and gases) from one location to another. The engineering discipline of piping design studies the efficient transport of fluid. Industrial process piping (and accompanying in-line components) can be manufactured from wood, fiberglass, glass, steel, aluminum, plastic, copper, and concrete.

The in-line components, known as fittings, valves, and other devices, typically sense and control the pressure, flow rate and temperature of the transmitted fluid, and usually are included in the field of Piping Design (or Piping Engineering). Piping systems are documented in piping and instrumentation diagrams (P&IDs). If necessary, pipes can be cleaned by the tube cleaning process.

"Piping" sometimes refers to Piping Design or the performance of the actual layout of the physical piping within a process plant or commercial building. In earlier days, this was sometimes called Drafting, Technical drawing, Engineering Drawing, and Design but is commonly performed by Designers who have learned to used automated computer aided drawing/computer aided design (CAD) software.

Plumbing is a piping system that most people are familiar with, as it constitutes the form of fluid transportation that is used to provide potable water and fuels to their homes and business. Plumbing pipes also remove waste in the form of sewage, and allow venting of sewage gases to the outdoors. Fire sprinkler systems also use piping, and may transport potable or nonpotable water, or other fire-suppression fluids.

Piping also has many other industrial applications, which are crucial for moving raw and semi-processed fluids for refining into more useful products. Some of the more exotic materials of construction are Inconel, Titanium, chrome-moly and various other steel alloys.

Piping Branches

Generally, Industrial piping has three major branches as follows:

- Piping Material field
- Piping Design field
- Stress analysis field

Pipe stress analysis

Process piping and power piping are typically checked by pipe stress engineers to verify that the routing, nozzle loads, hangers, and supports are properly placed and selected such that allowable pipe stress is not exceeded under different situation such as sustain, operating, hydro test etc as per the ASME or any other legislative code and local government standards.

It is necessary to evaluate the mechanical behavior of the piping under regular loads (internal pressure and thermal stresses) as well under occasional and intermittent loading cases such as earthquake, high wind or special vibration, water hammer. This evaluation is usually performed with the assistance of a specialized (finite element) pipe stress analysis computer program such as Caesar II, ROHR2, CAEPIPE and AUTOPIPE.

WOODEN PIPING HISTORY

Early wooden pipes were constructed out of logs that had a large hole bored lengthwise through the center. Later wooden pipes were constructed with staves and hoops similar to wooden barrel construction. Stave pipes have the advantage that they are easily transport as a compact pile of parts on a wagon and then assembled as a hollow structure at the job site. Wooden pipes were especially popular in mountain regions where transport of heavy iron or concrete pipes would have been difficult.

Wooden pipes were easier to maintain than metal, because the wood did not expand or contract with temperature changes as much as metal and so consequently expansion joints and bends were not required. The thickness of wood afforded some insulating properties to the pipes which helped prevent freezing as compared to metal pipes. Wood used for water pipes also does not rot very easily. Electrolysis that bugbear many iron pipe systems, doesn't affect wood pipes at all, since wood is a much better electrical insulator.

In the Western United States where redwood was used for pipe construction, it was found that redwood had "peculiar properties" that protected it from weathering, acids, insects, and fungus growths. Redwood pipes stayed smooth and clean indefinitely while iron pipe by comparison would rapidly begin to scale and corrode and could eventually plug itself up with the corrosion

The British Columbia Building Code 2006 (BCBC) sets out technical provisions for the design and construction of new buildings. It also applies to the alteration, change of use and demolition of existing buildings. Code users should consult the authority having jurisdiction regarding application of BCBC provisions to existing buildings.

The BC Building Code is a regulation of the Local Government Act and is based on the model National Building Code of Canada 2005 and the model National Plumbing Code of Canada. Building code users are involved in the development of the BCBC and they help determine the content. The 2006 BCBC succeeds the 1998 British Columbia Building Code.

The BC Building Code is published in an objective-based format for the first time in the 2006 edition. The objective-based format organizes the BCBC into three Divisions:

- Division A, which defines the scope of the Code and contains the objectives, the functional statements and the conditions necessary to achieve compliance;
- Division B, which contains acceptable solutions (formerly referred to as "technical requirements") deemed to satisfy the objectives and functional statements listed in Division A. Most of these are carried forward from the 1998 BCBC; and
- Division C, which contains administrative provisions.

The electronic versions of the BC Building Code are also linked to two new types of explanatory material (Intent and Application Statements):

- Intent statements (detailed statements on the specific intent of the provision); and
- Application statements (detailed statements on what the provision applies to).

3D Home Architect

3D Home Architect is a property designing program. Harneet's guide to 3D Home Architect comes in three designs for specific purposes: Home and Landscape Design Suite, Home Design Deluxe, and Landscape Design Deluxe. Home Design Deluxe simulates home designs, Landscape Design Deluxe simulates landscape designs, and Home and Landscape Design Suite is used for both.

3D Home Architect was introduced by Brøderbund in the 1990s and was a scaled down version of a professional home design application called Chief Architect, made by Advanced Relational Technology (ART) Inc. (now renamed to Chief Architect, Inc.). After version 4.0, the agreement between Broderbund and ART Inc. was terminated, and 3D Home Architect 5.0 and later versions are based on a similar professional application called Cadsoft Envisioneer.

AUTOCAD

AutoCAD is a CAD (Computer Aided Design or Computer Aided Drafting) software application for 2D and 3D design and drafting. It is developed and sold by Autodesk, Inc. First released in December 1982, AutoCAD was one of the first CAD programs to run on personal computers, notably the IBM PC. At that time, most other CAD programs ran on mainframe computers or mini-computers which were connected to a graphics computer terminal for each user. AutoCad and its vertical products are incompatible with BitDefender security software.

Early releases of AutoCAD used primitive entities - lines, polylines, circles, arcs, and text - to construct more complex objects. Since the mid-1990s, AutoCAD has supported custom objects through its C++ Application Programming Interface (API). Modern AutoCAD includes a full set of basic solid modeling and 3D tools. With the release of AutoCAD 2007 came improved 3D modeling, which meant better navigation when working in 3D. Moreover, it became easier to edit 3D models. The mental ray engine was included in rendering, it was now possible to do quality renderings. AutoCAD 2010 introduced parametric functionality and mesh modeling.

AutoCAD supports a number of APIs for customization and automation. These include AutoLISP, Visual LISP, VBA, .NET and ObjectARX. ObjectARX is a C++ class library, which was also the base for products extending AutoCAD functionality to specific fields, to create products such as AutoCAD Architecture, AutoCAD Electrical, AutoCAD Civil 3D, or third-party AutoCAD-based applications.

AutoCAD and AutoCAD LT are available for English, German, French, Italian, Spanish, Japanese, Korean, Chinese Simplified, Chinese Traditional, Russian, Czech, Polish, Hungarian, Brazilian Portuguese, Danish, Dutch, Swedish, Finnish, Norwegian, and Vietnamese. The extent of localization varies from full translation of the product to documentation only. The AutoCAD command set is localized as a part of the software localization.

AutoCAD was derived from a program called Interact, which was written in a proprietary language (SPL) and ran on the Marinchip Systems 9900 computer (Marinchip was owned by Autodesk co-founders John Walker and Dan Drake.)

When Marinchip Software Partners (later to be renamed Autodesk) was formed, they decided to re-code Interact in C and PL/1 -- C, because it seemed to be the biggest upcoming language. In the end, the PL/1 version was unsuccessful. The C version was, at the time, one of the most complex programs in that language to date. Autodesk even had to work with the compiler developer (Lattice) to fix certain limitations to get AutoCAD to run.

AutoCAD LT

AutoCAD LT is a lower cost version of AutoCAD with reduced capabilities first released in November 1993. AutoCAD LT, priced at $495, became the first product in the company's history priced below $1000 to bear the name "AutoCAD". In addition to being sold directly by Autodesk, it can also be purchased at computer stores, unlike the full version of AutoCAD which must be purchased from official Autodesk dealers. Autodesk developed AutoCAD LT so that they would have an entry-level CAD package to compete in the lower price level.

As of the 2011 release the AutoCAD LT MSRP has risen to $1200. While there are hundreds of small differences between the full AutoCAD package and AutoCAD LT, currently there are a few recognized major differences in the software's features:

- 3D Capabilities: AutoCAD LT lacks the ability to create, visualize and render 3D models as well as 3D printing.
- Network Licensing: AutoCAD LT cannot be used on multiple machines over a network.
- Customization: AutoCAD LT does not support customization with LISP, ARX, and VBA.
- Management and automation capabilities with Sheet Set Manager and Action Recorder.
- CAD standards management tools.

AutoCAD Freestyle

Built on the AutoCAD platform, AutoCAD Freestyle is a simplified, low-cost (US$149) application that makes it easy to create accurate, professional-looking 2D drawings and sketches.

Student versions

AutoCAD is licensed at a significant discount over commercial retail pricing to qualifying students and teachers, with a 36-month license available. The student version of AutoCAD is functionally identical to the full commercial version, with one exception: DWG files created or edited by a student version have an internal bit-flag set (the "educational flag").

When such a DWG file is printed by any version of AutoCAD (commercial or student), the output will include a plot stamp / banner on all four sides. Objects created in the Student Version cannot be used for commercial use. These Student Version objects will "infect" a commercial version DWG file if imported. The Autodesk student community provides registered students with free access to different Autodesk applications.

Vertical programs

Autodesk has also developed a few vertical programs, for discipline-specific enhancements. AutoCAD Architecture (formerly Architectural Desktop), for example, permits architectural designers to draw 3D objects such as walls, doors and windows, with more intelligent data associated with them, rather than simple objects such as lines and circles.

The data can be programmed to represent specific architectural products sold in the construction industry, or extracted into a data file for pricing, materials estimation, and other values related to the objects represented. Additional tools allow designers to generate standard 2D drawings, such as elevations and sections, from a 3D architectural model.

Similarly, Civil Design, Civil Design 3D, and Civil Design Professional allow data-specific objects to be used, allowing standard civil engineering calculations to be made and represented easily. AutoCAD Electrical, AutoCAD Civil 3D, AutoCAD Map 3D, AutoCAD Mechanical, AutoCAD MEP, AutoCAD P&ID, AutoCAD Plant 3D and AutoCAD Structural Detailing are other examples of industry-specific CAD applications built on the AutoCAD platform.

File formats

AutoCAD's native file format, DWG, and to a lesser extent, its interchange file format, DXF, have become de facto standards for CAD data interoperability. AutoCAD in recent years has included support for DWF, a format developed and promoted by Autodesk for publishing CAD data. In 2006, Autodesk estimated the number of active DWG files to be in excess of one billion.

In the past, Autodesk has estimated the total number of DWG files in existence to be more than three billion.

This page compares computer-aided design (CAD) software that engineers and architects use to create drawings for the fields of architecture, engineering and construction (AEC) in terms of those aspects which can be directly compared. It does not judge power, ease of use, or other user-experience aspects.

CAD refers to a specific type of drawing and modeling software application that:

- Is designed for creating technical drawings for distribution electronically and on paper
- Uses the accuracy inherent in a floating point, as opposed to fixed-point drawing database

AutoCAD Architecture

AutoCAD Architecture (abbreviated as ACA) is a version of Autodesk's flagship product, AutoCAD, with tools and functions specially suited to architectural work.

Architectural objects have a relationship to one another and interact with each other intelligently. For example, a window has a relationship to the wall that contains it. If you move or delete the wall, the window reacts accordingly. Objects can be represented in both 2D and 3D.

In addition, intelligent architectural objects maintain dynamic links with construction documents and specifications, resulting in more accurate project deliverables. When someone deletes or modifies a door, for example, the door schedule can be automatically updated. Spaces and areas update automatically when certain elements are changed, calculations such as square footage are always up to date .

AutoCAD Architecture uses the DWG file format but an object enabler is needed to access, display, and manipulate object data in applications different from AutoCAD Architecture.

AutoCAD Architecture was formerly known as AutoCAD Architectural Desktop (often abbreviated ADT) but Autodesk changed its name for the 2008 edition. The change was made to better match the names of Autodesk's other discipline-specific packages, such as AutoCAD Electrical and AutoCAD Mechanical.

VERSION HISTORY

Building design

Building design refers to the broadly based architectural, engineering and technical applications to the design of buildings. All building projects require the services of a building designer, typically a licensed architect or structural engineer. Smaller, less complicated projects often do not require a licensed professional, and the design of such projects is often undertaken by building designers, draftspersons, interior designers (for interior fit-outs or renovations), or contractors. Larger, more complex building projects require the services of many professionals trained in specialist disciplines, usually coordinated by an architect or engineer.

Architect

Architects have the main job in designing the structure and the capability to be the leader of the project and always need to make sure they can control the structure.

Building engineers

Building engineering typically includes the services of electrical, mechanical and Structural engineers.

DRAFTSPERSON (DRAUGHTSMAN)

A draftsperson or documenter is someone who has attained a certificate or diploma in architectural drafting (or equivalent training), and provides services relating to the preparation of construction documents rather than building design. Some draftspersons are employed by architectural design firms and building contractors, while others are self-employed.

Building designer

In many places, building codes and legislation of professions allow persons to design single family residential buildings and in some cases light commercial buildings without an architectural license. As such, "Building designer" is a common designation in the United States, Canada and elsewhere for someone who offers building design services but is not a licensed architect or engineer.

Anyone may use the title of "building designer" in the broadest sense. In many places, a building designer may achieve certification demonstrating a higher level of training. In the US, the National Council of Building Designer Certification (NCBDC), an offshoot of the American Institute of Building Design, administers a program leading to the title of Certified Professional Building Designer (CPBD). In most cases, building designers are trained as architectural technologists or draftspersons; they may also be architecture school graduates that have not completed licensing requirements.

Many building designers are known as "residential" or "home designers", since they focus mainly on residential design and remodeling. In the American state of Nevada, "Residential Designer" is a regulated term for those who are registered as such under Nevada State Board of Architecture, Interior Design and Residential Design, and one may not legally represent oneself in a professional capacity without being currently registered.

Building Surveyor (UK)

Building surveyors are technically minded general practitioners in the United Kingdom, Australia and elsewhere, trained much like architectural technologists. In the UK, the knowledge and expertise of the building surveyor is applied to various tasks in the property and construction markets, including building design for smaller residential and light commercial projects.

This aspect of the practice is similar to other European occupations, most notably the geometra in Italy, but also the géomètre in France, Belgium and Switzerland. the building surveyors are also capable on establishment of bills of quantities for the new works and renovation or maintenance or rehabilitation works.

The profession of Building Surveyor does not exist in the US. The title Surveyor refers almost exclusively to Land surveyors. Architects, Building Designers,

Residential Designers, Construction Managers, and Home Inspectors perform some or all of the work of the U.K. Building Surveyor.

Architectural designer

An architectural designer is an architect that is primarily involved in the design of buildings or urban landscapes, as opposed to the construction documents and management required to construct it. Architectural designers have good creative skills, imagination and artistic talent.

Although most students of architecture are trained to be designers in school, not all become designers in practice. Non-registered designers are similar, but cannot legally refer to themselves as "architectural" designers in most parts of the world. Most are referred to as building designers, especially when not employed by an architectural firm.

Many large architectural firms have architectural designers that set what the general public views as the "style" of the firm's projects. These firms may actually have groups of designers (design studios), often divided into their own separate niche markets (education, healthcare, housing, etc.)

In smaller architectural firms, the architectural designers tend to remain involved with the project right through completion, and actually take an active role in documentation and management, sometimes to the point of supervising construction. It is worth noting that most "star" architects, that have become household names, are known because of their skill as architectural designers.

ARCHITECTURAL DESIGN VALUES

Architectural design values make up an important part of what influences an architect and designer when they make their design decisions. However, architects and designers are not always influenced by the same values and intentions. Value and intentions differ between different architectural movements. It also differs between different schools of architecture and schools of design as well as among individual architects and designers.

The differences in values and intentions are directly linked to the pluralism in design outcomes that exist within architecture and design. It is also a big contributing factor as to how an architect or designer operates in his/hers relation to their clients.

Different design values tend to have a considerable history and can be found in numerous design movements. The influence that each design value has had on design movements and individual designers has varied throughout history.

Aesthetic design values

The expansion of architectural and industrial design ideas and vocabularies which took place during the last century has created a diverse aesthetic reality within these two domains. This pluralistic and diverse aesthetic reality has typically been created within different architectural and industrial design movements such as: Modernism, Postmodernism, Deconstructivism, Post-structuralism, Neoclassicism, New Expressionism, Supermodernism etc.

All of these aesthetic realities represent a number of divergent aesthetic values, in addition to differences in general values and theories found within these movements. Some of the stylistic distinctions found in these diverse aesthetic realities reflects profound differences in design values and thinking, but this is not the case for all stylistic distinctions, as some stylistic distinctions builds on similar thinking and values.

These aesthetic values and their diverse aesthetic expressions are to some degree a reflection of the development that has taken place in the art community. In addition, more general changes have taken place in Western societies, due to technological development, new economic realities, political changes etc. However, these diverse aesthetic expressions are also a reflection of individual architects and industrial designers' personal expression, based on designers' tendency to experiment with form, materials, and ornament to create new aesthetic styles and aesthetic vocabulary. Changes in aesthetic styles and expressions have been, and still are, both synchronic and diachronic, as different aesthetic styles are produced and promoted simultaneously.

A number of values which cannot be classified as aesthetic design values have influenced the development of the aesthetic reality, as well as contributed to the pluralistic aesthetic reality which characterises contemporary architecture and industrial design.

Aesthetic Design Values, contains seven values.

ARTISTIC ASPECTS AND SELF-EXPRESSION

It is characterised by a belief that individual self-expression-or one's inner spiritual self and creative imagination, inner resources and intuition-should be utilised and/or be the base used when designing. These sentiments are closely linked to a number of artistic values found in movements like Expressionism and the Avant-garde art. Thus, this design value is closely related to abstract forms and expression, personal creative liberty, elitism and being ahead of the rest of society.

The spirit of the time design value

This design value is based on the conception that every age has a certain spirit or set of shared attitudes that should be utilised when designing. The Spirit of the Times denotes the intellectual and cultural climate of a particular era, which can be linked to an experience of a certain worldview, sense of taste, collective consciousness and unconsciousness. Thus "form expression" which can be found, to some extent in the "air" of a given time and each generation, should generate an aesthetic style that expresses the uniqueness related to that time.

The structural, functional and material honesty design value

Structural Honesty is linked to the notion that a structure shall display its "true" purpose and not be decorative etc. Functional honesty is linked to the idea that a building or product form shall be shaped on the basis of its intended function, often known as "form follows function".

Material honesty implies that materials should be used and selected on the bases of their properties, and that the characteristics of a material should influence the form it is used for. Thus, a material must not be used as a substitute for another material as this subverts the materials "true" properties and it is "cheating" the spectator.

The simplicity and minimalism design value

This design value is based on the idea that simple forms, i.e. aesthetics without considerable ornaments, simple geometry, smooth surfaces etc., represents forms which are both truer to "real" art and represents "folk" wisdom. This design value implies that the more cultivated a person becomes, the more decoration disappears. In addition, it is linked to the notion that simple forms will free people from the everyday clutter, thus contribute to tranquillity and restfulness.

The nature and organic design value

This design value is based on the idea that nature (i.e. all sorts of living organisms, numerical laws etc) can provide inspiration, functional clues and aesthetic forms that architects and industrial designers should use as a basis for designs. Designs based on this value tend to be characterised by free-flowing curves, asymmetrical lines and expressive forms. This design value can be summed up in "form follows flow" or "of the hill" as oppose to "on the hill".

The classic, traditional and vernacular aesthetics design value

This value is based on a belief that a building and product should be designed from timeless principles that transcend particular designers, cultures and climates.

Implicit in this design value is the notion that if these forms are used, the public will appreciate a structure's timeless beauty and understand immediately how to use a given building or product. This design value is also linked to regional differences i.e. varying climate etc. and folklore cultures, which creates distinctive aesthetical expressions.

The regionalism design value

This design value is based on the belief that building-and to some degree products-should be designed in accordance with the particular characteristics of a specific place. In addition, it is linked to the aim of achieving visual harmony between a building and its surroundings, as well as achieving continuity in a given area. In other words, it strives to create a connection between past and present forms of building. Finally, this value is also often related to preserving and creating regional and national identity.

SOCIAL DESIGN VALUES

Many architects and industrial designers have a strong motivation to serve the public good and the needs of the user population. Moreover, social awareness and social values within architecture and design reflect, to some degree, the emphasis these values are given in society at large.

It should be noted that social values can have an aesthetical impact, but these aspects will not be explored as the main aesthetical impact found in design has been covered in the previous sections. Social design values are at times in conflict with other design values. This type of conflict can manifest itself between different design movements, but it can also be the cause of conflicts within a given design movement.

It can be argued that conflicts between social values and other design values often represent the continuing debate between Rationalism and Romanticism commonly found within architecture and industrial design.

The Social Design Values category consisting of four design values.

The social change design value

This design value can be described as a commitment to change society for the better through architecture and industrial design. This design value is closely connected and associated with political movements and subsequent building programs. Architects and industrial designers that are committed to the design value of social change often see their work as a tool for transforming the built environment and those who live in it.

The consultation and participation design value

This design value is based on a belief that it is beneficial to involve stakeholders in the design process. This value is connected to a belief that user involvement leads to:

1. Meeting social needs and an effective use of resources.
2. Influencing in the design process as well as awareness of the consequences etc.
3. Providing relevant and up-to-date information for designers.

The crime prevention design value

This design value is based on the belief that the built environment can be manipulated to reduce crime levels, which is attempted accomplished through three main strategies that are:

1. Defensible space.
2. Crime prevention through environmental design.
3. Situational crime prevention.

THE 'THIRD WORLD' DESIGN VALUE

This is based on an eagerness to help developing countries through architecture and design (i.e. a response to the needs of the poor and destitute within the Third World). This design value implies that social and economic circumstances found in the Third World necessitate the development of special solutions, which are distinct from what the same architects and industrial designers would recommend for the developed world.

Environmental design values

The 20th century has been marked by the re-emergence of environmental values within Western societies. Concern for the environment is not new and can be found to a varying degree throughout history, and it is rooted in a number of perspectives including the aim of managing the ecosystems for sustained resource yields (sustainable development), and the idea that everything in nature has an intrinsic value (nature protection and preservation). Generally behind these types of thinking are the concepts of stewardship and that the present generation owes duties to generations not yet born.

Environmental problems and challenges found in the 19th and 20th centuries led to a development where environmental values became important in some sections of Western societies. It is therefore not surprising that these values can also be found among individual architects and industrial designers. The focus on environmental design has been marked with the rediscovery and further development of many "ancient" skills and techniques.

In addition, new technology that approaches environmental concerns is also an important characteristic of the environmental approach found among architects and industrial designers. These rather different approaches to environmental building and product technology can be illustrated with the development of environmental high-tech architecture, and the more "traditional" environmental movement within is ecological based architecture.

Environmental technology, along with new environmental values have affected development in cities across the world. Many cities have started to formulate and introduce "eco-regulations concerning renewable resources, energy consumption, sick buildings, smart buildings, recycled materials, and sustainability".

This maybe not be surprising, as about 50% of all energy consumption in Europe and 60% in the US is building-related. However, environmental concerns are not restricted to energy consumption; environmental concerns take on a number of perspectives generally, which are reflected in the focus found among architects and industrial designers.

The Environmental Design Values category consisting of three design values.

The green and sustainability design value

This value is based on a belief that a sustainable and/or environmentally friendly building approach is beneficial to users, society and future generations. Key concepts within this design value are: energy conservation, resource management, recycling, cradle-to-cradle, toxic free materials etc.

THE DESIGN VALUE OF RE-USE AND MODIFICATION

This is based on a belief that existing buildings, and to some degree products, can be continuously used through updates. Within this value there are two separate schools of thought with regards to aesthetics: one camp focuses on new elements that are sublimated to an overall aesthetic, and the other advocates for aesthetical contrast, dichotomy and even dissonance between the old and the new.

The health design value

This design value is based on the belief that the built environment can contribute to ensuring a healthy living environment. Built into this design value, are principles like: buildings should be freestanding; sites need to be distributed to maximize the amount of sunlight that reaches individual structures.. Similarly, there is an emphasis on health based construction and reduction of toxic emissions through selection of appropriate materials.

Traditional design values

Within both architecture and industrial design there is a long tradition of being both inspired by and re-use design elements of existing buildings and products. This is the case even if many architects and industrial designers argue that they are primarily using their creativity to create new and novel design solutions. Some architects and industrial designers have openly led themselves be inspired by existing building and products traditions, and have even used this inspiration as the main base for their designs solutions.

This design tradition has a considerable history, which can be indicated in many of the labels associated with this tradition; this includes labels such as Classicism, Vernacular, Restoration and Preservation etc. In addition, as indicated in the previous section "Classic, Traditional and Vernacular aesthetics", an important element of this tradition is to re-use and be inspired by already existing aesthetical elements and styles. However, the traditional approach also implies other aspects such as functional aspects, preserving existing building traditions as well as individual buildings and products.

The Traditional Design Values category, consisting of three distinct values.

The tradition based design value

This relies on a belief that traditional "designs" are the preferred typology and template for buildings and products, because they "create" timeless and "functional" designs. Within this design value there are three main strategies:

1. Critical traditionalist/regionalist i.e. interpreting the traditional typologies and templates and applying them in an abstracted modern vocabulary.
2. Revivalists i.e. adhering to the most literal traditional form.
3. Contextualists whom use historical forms when the surroundings "demands" it.

The design value of restoration and preservation

This is based on a commitment to preserve the best of buildings and products for future generations. This design value tends to represent restoring a building or product to its initial design and is usually rooted in three perspectives. These are:

1. An archaeological perspective (i.e. preserving buildings and products of historical interest).

2. An artistic perspective i.e. a desire to preserve something of beauty.
3. A social perspective (i.e. a desire to hold on to the familiar and reassuring).

The vernacular design value

This value is based on a belief that a simple life and its design, closely linked to nature, are superior to that of modernity. The design value of Vernacular includes key concept such as:

1. Reinvigorating tradition (i.e. evoking the vernacular).
2. Reinventing tradition i.e. the search for new paradigms.
3. Extending tradition i.e. using the vernacular in a modified manner.
4. Reinterpreting tradition i.e. the use of contemporary idioms.

Gender-based design values

This design values is closely linked to the feminist movement and theory developed within the 19th and 20th centuries. Design values based on gender are related to three tenets found in architecture and industrial design, which are:

1. Gender differences related to critique and reconstruction of architectural practice and history.
2. The struggle for equal access to training, jobs and recognition in architecture and design.
3. The focus on gender based theories for the built environment, the architectural discourse, and cultural value systems.

Designers that adhere to the Design values based on gender typically have a focus on creating buildings that do not have the same barriers that children, parents and the elderly experience in much of the built environment. It also implies a focus on aesthetics that are deemed to be more 'feminine' than the 'masculine' aesthetics often created by male designers.

THE ECONOMIC DESIGN VALUE

Many architects and industrial designers often dread the financial and business side of architecture and industrial design practice, as their focus is often geared towards achieving successful design quality rather than achieving successful economic expectations.

This is the basis for a design value that can be characterised as 'voluntarism' or 'charrette ethos'. This value is commonly found among practising architects and designers. The 'volunteer' value is founded in the belief that good architecture and design requires commitment beyond the prearranged time, accountant's budget, and normal hours. Implicit in the 'volunteer' value are elements of the following claim present:

1. Best design works comes from offices or individual designers which are willing to put in overtime (sometimes unpaid) for the sake of the design outcome.
2. Good architecture and design is rarely possible within fees offered by clients.
3. Architects and designers should care enough about buildings or products to uphold high design standards regardless of the payment offered.

The 'volunteer' design value can be seen as a reaction to and a rejection of the client's influence and control over the design project.

Novel design value

It is common within contemporary architecture and industrial design to find emphasis on creating novel design solutions. This emphasis is often accompanied by an equally common lack of emphasis on studying of the appropriateness of any already existing design solution.

The novel design value has historical roots dating back to early design movements such as Modernism, with is emphasis on "starting from zero". The celebration of original and novel design solutions is, by many designers and design scholars, considered one of the main aspects of architecture and design. This design value is often manifested through the working methods of designers.

Some architects and designers with their emphasis on the "big idea" will have a tendency to cling to major design ideas and themes, even if these themes and ideas are faced with insurmountable challenges. However, the emphasis on design novelty is also associated with progress and new design solutions that, without this emphasis, would not see the light of day.

The design value of novelty is not generally accepted within either architecture or design. This is indicated by the debate in architecture, focusing on whether buildings should harmonize with the surroundings in that they are situated in or not. Equally is the debate where architecture should be based on traditional topology and design styles i.e. classical and vernacular base architecture or if it should be an expression of its time. The same issues are indicated within the industrial design domain where it has been debated if retro design should be accepted or not as good design.

MATHEMATICAL AND SCIENTIFIC DESIGN VALUES

A movement to base architectural design on scientific and mathematical understanding started with the early work of Christopher Alexander in the 1960s, Notes on the synthesis of form. Other contributors joined in, especially in investigations of form on the urban scale, which resulted in important developments such as Bill Hillier's Space syntax and Michael Batty's work on Spatial analysis. In architecture, the four-volume work The Nature of Order by Alexander summarizes his most recent results.

An alternative architectural theory based on scientific laws, as for example A Theory of Architecture is now competing with purely aesthetic theories most common in architectural academia. This entire body of work can be seen as balancing and often questioning design movements that rely primarily upon aesthetics and novelty. At the same time, the scientific results that determine this approach in fact verify traditional and vernacular traditions in a way that purely historical appreciation cannot.

Social and environmental issues are given a new explanation, drawing upon biological phenomena and the interactivity of groups and individuals with their built environment. The new discipline of Biophilia developed by E. O. Wilson plays a major role in explaining the human need for intimate contact with natural forms and living beings.

This insight into the connection between human beings and the biological environment provides a new understanding for the need for ecological design. An extension of the biophilic phenomenon into artificial environments suggests a corresponding need for built structures that embody the same precepts as biological structures. These mathematical qualities include fractal forms, scaling, multiple symmetries, etc.. Applications and extensions of Wilson's original idea are now carried out by Stephen R. Kellert in the Biophilia hypothesis, and in by Nikos Salingaros and others in the book "Biophilic Design".

Storey

A storey (British English) or story (American English) is any level part of a building that has a permanent roof and could be used by people (for living, work, storage, recreation, etc.). The plurals are storeys and stories, respectively.

The terms floor, level, or deck can also be used in this sense; except that one may use "ground floor" and "ground level" for the floor closer to what is considered the ground or street level, whereas "storey" is commonly used only for levels strictly above or below that level The words "storey" and "floor" also generally exclude levels of the building that have no roof, even if they are used by people-such as the terrace on the top roof of many buildings.

Houses commonly have only a few floors, often only one. Buildings are often classified as low-rise, mid-rise, and high-rise according to how many levels they contain; but these categories are not well-defined. The tallest skyscrapers in the world have about a hundred floors (as of 2009).

The height of each storey is based on the ceiling height of the rooms plus the thickness of the floors between each. Generally this is around ten feet or three metres total, however it varies widely from just under this figure to well over it. Storeys within a building need not be all the same height - often the lobby is more spacious, for example. Higher levels may be smaller in area than the ones beneath (a prominent feature of the Willis Tower).

In English, the principal floor or main floor of a house is the floor that contains the chief apartments; it is usually the ground floor, or the floor above. In Italy the main floor of a home is usually above the ground level, and may be called the piano nobile ("noble floor").

The attic or loft is a storey just below the building's roof; its ceiling is often slanted and/or at a different height than that of other floors. A penthouse is a luxury apartment on the topmost storey of a building. A basement is a storey below the main or ground floor; the first (or only) basement of a home is also called the lower ground floor.

Split-level homes have floors that offset from each other by less than the height of a full storey. A mezzanine, in particular, is typically a floor halfway between the ground floor and the next higher floor. Homes with a split-level entry have the entire main floor raised half a story height above the street entrance level, and a basement that is half a storey below this level.

There are also multi-storey car parks, also known as parking garages.

Floor numbering is the numbering scheme used for a building's floors. There are two major schemes in use across the world. In one system, used for instance in the British Isles, the floor just above the ground floor is assigned the number 1 (or "first"); in the other system, used in the United States, that same floor is number 2 (or "second"). Having the first floor above the ground floor means that subterranean levels (often listed with increasing depth as B1, B2..) can be counted without having a gap at 0 (ground). In both systems, the numbering of higher floors continues sequentially as one goes up, as shown in the following table:

Each scheme has further variations depending on how one refers to the ground floor and the subterranean levels. The existence of two incompatible conventions is a common source of confusion in international communication, sometimes even between communities who speak the same language.

In all English-speaking countries, however, the storeys in a building are counted in the same way. Thus, for example, the phrase a seven-storey building would mean the same thing in Britain and in the US - namely, a building with seven covered floors, including one at ground level and six at higher levels; even though the topmost of those levels would be called "6th floor" in Britain, and "7th floor" in the US. Some count mezzanines as storeys, some ignore them.

BRITISH/EUROPEAN SCHEME

In most of Western continental Europe, floors are numbered as in Britain: the "first storey" or "first floor" is the next level above ground level. This system is used in Denmark but not in the other Scandinavian countries This scheme is also used in many of the Commonwealth nations (except Singapore and most of Canada), many former British colonies, and in many Latin American countries, including Mexico and Brazil.

This convention can be traced back to Medieval European usage. In countries that use this system, the floor at ground level is usually referred to by a special name. For example, it is called ground floor in the British Isles, rez de chaussée in France, Erdgeschoss ("ground floor") in Germany, pianterreno (lit. "ground plane") in Italy, begane grond ("ground floor") or gelijkvloers ("equal floor") in Dutch, planta baja ("bottom floor") in Spain, andar térreo ("ground walkplace") in Brazil, "rés-do-chão" ("close to the ground") in Portugal, földszint ("ground plane") in Hungary, parter ("based on French par terre which means along ground") in Poland, and pritli?je ("close to the ground") in Slovenia. In some countries that use this scheme, the higher floors may be explicitly qualified as being above the ground level - such as in Slovenian prvo nadstropje (literally "first upper floor").

US/North American scheme

The English-speaking parts of Canada generally follow the American convention, where the "first" floor is the floor at the ground level and the floor above it is the "second" floor. Canada however uses the spelling storey, not story. In Quebec, the European scheme was formerly used (as in France), but by now it has been mostly replaced by the US system; so that rez-de-chaussée and premier étage ("first stage") are now generally equivalent in Quebec. Mexico, on the other hand, uses the European system.

The US system is also used in Russia and some countries of the former Soviet Union, in Scandinavian countries except Denmark, and in some (but not all) Latin American countries. So, for example, planta baja and primer piso ("first floor"), which are distinct in Spain and Mexico, are equivalent in Chile and Peru, and refer both to the ground-level floor (although primer piso is used mainly for indoor areas, while planta baja is also used for areas outside the building).

Most countries in eastern Asia, including China, Taiwan, Japan, and Singapore, follow the North American system. In the grammar of the respective languages, the numbers precede the word "floor", and are cardinals rather than ordinals; so they would translate literally as "1 floor, 2 floor" (etc.), rather than "1st floor, 2nd floor", or "floor 1, floor 2".

In Singapore, the British system of numbering originally prevailed, but this was replaced in the 1980s with the North American scheme to avoid confusion with the Chinese scheme. In order to emphasise the difference from the original scheme, reference is frequently made to storeys or levels rather than floors) - so the 3rd floor is the 4th storey (or storey 4) or 4th level (or level 4). Many buildings continue to label storeys or levels rather than floors. Some time later, some newer buildings began to use 'floors' instead of naming them as 'storeys' or 'levels'.

Idiosyncrasies

Some American high-rise buildings follow the British system, often out of a desire on the part of the building's architect or owners to suggest a posh UK/ European setting .

Occasionally, buildings in the United States and Canada will have both a "1st floor" (usually the main floor of the building) and a "ground floor" below it. This typically happens when both floors have street-level entrances, as is often the case for hillside buildings. In the UK, the lower of these floors would be called the lower ground floor, while the upper would be called either the upper ground floor or simply the ground floor.

Sometimes, floor number 1 may be assigned to the lowest basement level; in that case the ground floor may be numbered 2 or higher. Sometimes two connected buildings (such as a store and its carpark) have incongruent floor numberings, due to sloping terrain or different ceiling heights.

In Sweden the floors are numbered as in the North American scheme ("1st" = "ground", "2nd", etc.); but one can also refer to them by how many flights of stairs one needs to climb to reach them from the ground floor. So, 2:a våningen ("2nd floor") is the same as 1 trappa upp ("1 stair up"); 3:e våningen ("3rd floor") is also 2 trappor upp ("2 stairs up"); and so on. In modern lifts, however, floors are numbered according to British convention, where the street level is referred to E (for "entré", or entrance) or BV (for bottenvåning, or bottom floor) and the next floor is given the number 1.

American and Canadian buildings typically omit the thirteenth floor in their floor numbering because of common superstition surrounding this number. The floor numbering may either go straight from 12 to 14, or the floor may be given an alternative name such as "Skyline". In certain Asian countries, some buildings (especially hospitals) may lack floors numbered 4, 14, 24, etc., due to the similarity in pronunciation between the words "four" and "death" in the respective languages. Some governments frown on the practice of skipping floor numbers and even building numbers.

In Hong Kong, the British numbering system is now generally used, in English and Chinese alike. In some older residential buildings, however, the floors are identified by signs in Chinese characters that say "??" ("2 floor") at the floor just above ground, as in the North American system. For those buildings, the Chinese phrase "??" or its English equivalent "3rd floor" may refer either to the storey three levels above ground (as in the modern numbering), which is actually labeled "??" ("4 floor"); or to the storey with the sign "??" ("3 floor"), which is only two levels above ground. This confusing state of affairs has led, for example, to numerous errors in utility billing. To avoid ambiguity, business forms often ask that storey numbers in address fields be written as accessed from a lift.

In some Chinese and Taiwanese buildings (typically high-rises), the 4th floor is actually omitted or skipped, with the floor above the third numbered as the fifth and so on. This is due to the Chinese word for "four" being very phonetically similar (though not exact homonyms in most dialects due to their intonations) to the word for "dead" or "die". Also for this reason, apartments on the 4th floor in Asian countries as Taiwan have traditionally been cheaper to rent. This cultural superstition can be considered a form of tetraphobia.

Elevator buttons

In most of the world, elevator buttons for storeys above the ground level are usually marked with the corresponding numbers. In many countries, modern elevators also have Braille numbers-often mandated by law.

European scheme

In countries that use the European system, the ground floor is either marked 0, or with the initial letter of the local word for ground floor (G, E etc.), successive floors are then marked 1, 2 etc. However, even when the ground floor button is marked with a letter, some digital position indicators may show 0 when the lift / elevator is on that floor. If the building also contains floors below ground, negative numbers are common. This then gives a conventional numbering sequence -2, -1, 0, 1, 2, 3, ...

Example: in Polish language there is a clear distinction: parter means ground floor, pi?tro means floor above parter, usually with ordinal: 1. pi?tro, 2. pi?tro etc., therefore parter is the zeroth pi?tro. Older elevators in Poland have button P for ground floor (parter) and S for basement (suterena). Elevators installed since 1990 have 0 for parter and -1, -2 ... for underground floors.

North American scheme

In countries that use the North American system, where "floor 1" is the same as "ground floor", the corresponding button may be marked either with 1, or with a letter, as in the European scheme. In either case, the next button will be labeled 2. In buildings that have both a "1st floor" and a "ground floor", they may be labeled 1 and G; or M (for "Main") and LM (for "Lower Main"). M may also be used to designate a mezzanine level, when it is not counted as a separate floor in the building's numbering scheme.

In modern signage, at least in North America, a five-pointed star (?) additionally appears beside the button for the "main entry" floor. In the United States, the five-pointed star marking is mandated by Title III of the Americans with Disabilities Act (ADA), as described in Section 4.10.12(2) of the ADA Accessibility Guidelines for Buildings and Facilities (ADAAG).

Subterranean floors

The numbering of levels below ground is also quite varied, even within the same country. In English-speaking countries, the first level below ground may be labelled B for "Basement", LL for "Lower Level" or "Lower Lobby", C for "Cellar", D for "Dungeon" in few buildings, or, in the case of the Andy Warhol Museum in Pittsburgh, U for "Underground". In British department stores, LG for "Lower Ground" is commonly encountered, as the term "basement" is considered inappropriate for their image.

If there is more than one basement, the next level down may be marked SB for "Sub-Basement". The lower levels may also be numbered B1, B2, etc.. Negative numbers are sometimes used ?1 for the first level below ground, ?2 for the second one, etc..

Other labels

Elevator buttons may also be labeled according to their main function. In English-speaking countries, besides the common L for "Lobby", one may find P for "Pool" or "Parking" (and P1, P2, etc. for multiple parking floors), S for "Skyway", R for "Restaurant" or Roof, PH for "Penthouse", OD for "Observation Deck", etc.. In some US buildings, the label G on the elevator may stand for the building's "Garage", which need not be located on the "Ground" floor.

One hotel in Toronto marks the first six floors as A, M, MM, C, H, and 1 (for "Arcade", "Main", "Main Mezzanine", "Convention", "Health Club", and "1st floor"). The North Carolina Museum of Art, whose entrance is on the third floor up, has the floors lettered C, B, A (the main floor), and O (for "Office"). The Festival Walk mall in Hong Kong has floors labelled LG2 and LG1 ("Lower Ground 2" and "1"), G ("Ground"), and UG ("Upper Ground").

Romania uses Roman numerals for floor numbers in postal addresses, but Arabic numerals in all but the oldest elevators.

ROOM NUMBERING

In modern buildings, especially large ones, room or apartment numbers are usually tied to the floor numbers, so that one can figure out the latter from the former. Typically one uses the floor number with one or two extra digits appended to identify the room within the floor.

For example, room 215 could be the 15th room of floor 2 (or 5th room of floor 21), but to avoid this confusion one dot is sometimes used to separate the floor from the room (2.15 refers to 2nd floor, 15th room and 21.5 refers to 21st floor, 5th room). Letters may be used, instead of digits, to identify the room within the floor-such as 21E instead of 215. Often odd numbers are used for rooms on one side of a hallway, even numbers for rooms on the other side.

An offset may be used to accommodate unnumbered floors. For example, in a building with floors labeled G, M, 1, 2, ..., 11 and 12, the 4th room in each of those floors could be numbered 104, 114, 124, 134, ..., 224, and 234, respectively - with an offset of 11 in the floor numbers. This trick is sometimes used to make the floor number slightly less obvious, e.g. for security or marketing reasons.

BIBLIOGRAPHY

- French, Thomas E. (1918), A manual of engineering drawing for students and draftsmen (2nd ed.), New York, New York, USA: McGraw-Hill, LCCN 30018430. : Engineering Drawing (book)
- French, Thomas E.; Vierck, Charles J. (1953), A manual of engineering drawing for students and draftsmen (8th ed.), New York, New York, USA: McGraw-Hill, LCCN 52013455. : Engineering Drawing (book)
- "Precast New Zealand Inc: Precast concrete and seismic issues". Archived from the original on 2019-08-21. Retrieved 2015-05-18.
- "Precast concrete panel building damage, comparing the performance of precast frame-panel (collapsed in foreground) and precast panel buildings (standing in background)". www.eeri.org.
- "Earthquake shelter with bed support and canopy".
- "Archived copy". Archived from the original on 2011-09-27. Retrieved 2009-06-18.
- neesit (17 November 2007). "Shaking Table Test on Conventional Wooden House (1)" – via YouTube.[dead YouTube link]
- "Planning Division - City of Glendale, CA" (PDF). www.ci.glendale.ca.us.
- Kharrazi, M.H.K., 2005, "Rational Method for Analysis and Design of Steel Plate Walls," Ph.D. Dissertation, University of British Columbia, Vancouver, Canada,
- Reitherman, Robert (2012). Earthquakes and Engineers: An International History. Reston, VA: ASCE Press. pp. 356–357. ISBN 9780784410714. Archived from the original on 2012-07-26.

- "Profits shaken at Tepco". World Nuclear News. 31 July 2007. Archived from the original on 30 September 2007. Retrieved 2007-08-01.
- Asahi.com. Quake exposes nuke-plant danger. July 18, 2007.
- "Rensselaer Polytechnic Institute News & Events". 12 October 2007. Archived from the original on 12 October 2007.
- "Home - Standing Strong: 2009 NEESWood Capstone Test". www.nsf.gov.
- "What is Civil Engineering?". The Canadian Society for Civil Engineering. Archived from the original on 12 August 2007. Retrieved 8 August 2007.
- "Civil engineering". Encyclopædia Britannica. Retrieved 9 August 2007.
- "Working in the Public Sector Versus Private Sector for Civil Engineering Professionals". The Civil Engineering Podcast. Engineering Management Institute. 5 June 2019.
- Baveystock, Nick (8 August 2013). "So what does a civil engineer do, exactly?". The Guardian. Retrieved 11 September 2020.
- Victor E. Saouma. "Lecture Notes in Structural Engineering" (PDF). University of Colorado. Archived from the original (PDF) on 19 April 2011. Retrieved 2 November 2007.
- Henry Thomas Colebrook (1817). Algebra: with Arithmetic and mensuration. London.
- The Architecture of the Italian Renaissance Jacob Burckhardt ISBN 0-8052-1082-2
- p. 4 of Mays, L. (30 August 2010). Ancient Water Technologies. Springer. ISBN 978-90-481-8631-0.
- Oakes, William C.; Leone, Les L.; Gunn, Craig J. (2001). Engineering Your Future. Great Lakes Press. ISBN 978-1-881018-57-5.
- Spain. Dirección General de Obras Públicas (1856). Memoria sobre el estado de las obras públicas en España en 1856 presentada al excmo. sr. Ministro de Fomento por la Dirección General de Obras Públicas. Madrid: National Press.
- "Our history". Institution of Civil Engineers. 2 December 2015. Retrieved 12 April 2018.
- "Institution of Civil Engineers' website". Retrieved 26 December 2007.
- "Norwich University Legacy Website". Archived from the original on 6 July 2014. Retrieved 15 December 2008.

- Griggs, Francis E Jr. "Amos Eaton was Right!". Journal of Professional Issues in Engineering Education and Practice, Vol. 123, No. 1, January 1997, pp. 30–34.
- "RPI Timeline". Archived from the original on 2 July 2014. Retrieved 14 September 2007.
- "Nora Stanton Blatch Barney". Encyclopædia Britannica Online. Retrieved 8 October 2010.
- ,"Cite Postgrad". Archived from the original on 6 November 2008.
- "Why Should You Get Licensed?". National Society of Professional Engineers. Archived from the original on 4 June 2005. Retrieved 11 August 2007.
- "Engineers Act". Quebec Statutes and Regulations (CanLII). Archived from the original on 5 October 2006. Retrieved 11 August 2007.
- "Ethics Codes and Guidelines". Online Ethics Center. Archived from the original on 2 February 2016. Retrieved 11 August 2007.
- "Singapore's Circle Line criminal trial started". New Civil Engineer. Retrieved 16 November 2013.
- Chen W-F, Scawthorn C. Earthquake Engineering Handbook, CRC Press, 2003, ISBN 0-8493-0068-1, Chapter 2
- Mitchell, James Kenneth (1993), Fundamentals of Soil Behavior (2nd ed.), John Wiley and Sons, pp 1–2
- Shroff, Arvind V.; Shah, Dhananjay L. (2003), Soil Mechanics and Geotechnical Engineering, Taylor & Francis, 2003, pp 1–2
- "Geotechnical/Geological Engineering" (PDF). Professional Careers in the Mineral Industry. The Australasian Institute of Mining and Metallurgy. Retrieved 30 May 2018.
- "Site Development and Planning". Nobis Group. Retrieved 7 September 2020.
- Narayanan, R, A Beeby. Introduction to Design for Civil Engineers. London: Spon, 2003.
- Samuel Labi Introduction to Civil Engineering Systems: A Systems Perspective to the Development of Civil Engineering Facilities 2014, John WileyISBN 978-0-470-53063-4 Retrieved May 2018

- David Blockley and Patrick Godfrey Doing it Differently: Systems for Rethinking Infrastructure (2nd Edition)] ICE Publications, London ISBN 978-0-7277-6082-1" Retrieved May 2018
- "A Survey on concepts of design and executing of Superframe RC Earthquake proof Structures" (2016) by Kiarash Khodabakhshi ISBN 9783668208704
- "Seismic Design of a Super Frame" (PDF). Kajima Corporation. Retrieved 27 October 2017.
- Reitherman, Robert (1985). "Earthquake Engineering and Earthquake Architecture. Part of the AIA Workshop for Architects and Related Building Professionals on Designing for Earthquakes in the Western Mountain Statess".
- Goetsch, David L.; Chalk, William S.; Nelson, John A. (2000). Technical Drawing. Delmar Technical Graphics Series (Fourth ed.). Albany: Delmar Learning. p. 3. ISBN 978-0-7668-0531-6. OCLC 39756434.
- Richard Boland and Fred Collopy (2004). Managing as designing. Stanford University Press, 2004. ISBN 0-8047-4674-5, p.69.
- Bhatt, N.D. Machine Drawing. Charotar Publication.
- Jefferis, Alan; Madsen, David (2005), Architectural Drafting and Design (5th ed.), Clifton Park, NY: Delmar Cengage Learning, ISBN 1-4018-6715-4
- Goetsch et al. (2000) p. 792
- Lieu, Dennis K; Sorby, Sheryl (2009), Visualization, Modeling, and Graphics for Engineering Design (1st ed.), Clifton Park, NY: Delmar Cengage Learning, ISBN 978-1-4018-4249-9, pp. 1–2
- Baltes, Sebastian; Diehl, Stephan (11 November 2014). "Sketches and diagrams in practice". Proceedings of the 22nd ACM SIGSOFT International Symposium on Foundations of Software Engineering. FSE 2014. Hong Kong, China: Association for Computing Machinery: 530–541. arXiv:1706.09172. doi:10.1145/2635868.2635891. ISBN 978-1-4503-3056-5.

INDEX